BEYOND EMPIRE AND REVOLUTION

BEYOND EMPIRE AND REVOLUTION

Militarization and Consolidation in the Third World

Irving Louis Horowitz

New York Oxford
OXFORD UNIVERSITY PRESS
1982

Library of Congress Cataloging in Publication Data

Horowitz, Irving Louis.
Beyond empire and revolution.

Includes index.
1. Underdeveloped areas—Economic policy.
2. Underdeveloped areas—Politics and government.
3. Underdeveloped areas—Social policy. I. Title.
HC59.7.H67 338.9'009172'4 80-27370
ISBN 0-19-502931-3 AACR1
ISBN 0-19-502932-1 (pbk.)

Printing (last digit): 9 8 7 6 5 4 3 2 1

Printed in the United States of America

To Mary

who has earned every
square inch of splendid
isolation afforded on
this dedication page

Preface

There is a perfectly sound basis for the convention of putting the Preface in roman numerals and the text in arabic numerals: namely, the presupposition that the Preface is written last. In my case at least, not until the final piece of this puzzle was put in place did I fully realize that this book might well have been entitled "Contra Extremis"—not in the sense of any fatuous search for a middle way or an ideological high ground, but in a paradigmatic sense that every chapter embodies a variation on a theme. That theme is the essential integrity and autonomy of the Third World, along with a leitmotif: those who assert that the Third World is simply on a developmental path toward modernization in accord with the First World or socialization in accord with the Second World are badly mistaken on empirical grounds and sadly misinformed on intellectual and moral grounds.

In the field of developmental studies, reputations have too often been built on the shifting sands of ideological style. When ours was heralded as the "American Century" one would have imagined that all roads led to a democratic/capitalist Mecca. When things turned sour, and American involvement in Vietnam helped turn the American Century into an American nightmare, the United States became, for many, the source of all evil with the same fervor of belief as previously it had been thought to be the source of all good.

Manicheanism replaced celebrations, and in all of this, the Third World became a sort of backdrop to events in the halls of power. The Third World became an object of strategy rather than stricture, subject to the policies of big powers instead of playing the politics of new nations. This book is intended to restore a sense of perspective to Third World studies.

The appropriate title for this volume, a matter of no small concern to any author, plagued me from the outset. My first impulse was to call the volume "Further Studies in Three Worlds of Development," since many of the essays derive from themes first enunciated in that earlier work. My work in international affairs has been marked by concern with the First World of capitalism, democracy, and the West as well as the Second World of socialism, authoritarianism, and the East, and how the strategies and tactics of development each world offers affect nations of the Third World individually and as a whole. The difficulty with this title, which ultimately ruled it out, is that events of the past decade have, in part, moved my analysis beyond the themes of that earlier work. International politics now concern a world of raw materials and foodstuffs, a world of neo-Malthusian dimensions, which looks and feels much different from the neo-Keynesian universe of the sixties, in which it was assumed that basic needs could be met, and thus the problems and processes of allocation and decision were paramount. This overwhelming shift in mood, if not always in substance, argued persuasively against a title that suggests nothing more than the extension of old formulations.

My second choice for a title was "Beyond the Third World." Certainly this would be a gripping title from the vantage point of the marketplace. The difficulty is that every page of my work reflects the reverse premise: that the Third World is undergoing a period of consolidation and structural unification. What might earlier have been arguable premises and tendencies have in the 1980s become full-scale realities and fully evolved trends. To have chosen a title that strongly implies the dissolution of the Third World, or the existence of a world configuration quite beyond the Third World, incorporating perhaps a Fourth World of starving, energy-poor nations, would have misrepresented this work. A Fourth World, however distinct, is a byproduct of Third World maturation. Yet, in a certain sense "Beyond the Third World" might have been an appropriate title: my analysis continually takes into account the triadic nature of world forces, analyzing the Third World in the context of a First World and Second World struggle, which I see as the essential source of the unstable equilibrium that now exists between nations, regions, and blocs. In this sense, my work does seek to move "Beyond the Third World." But this is such an unusual sense of the phrase that I have become persuaded that this title would have conveyed a profound injustice to the thrust and sense of my efforts.

However, if it would be inappropriate to entitle this book "Beyond the Third World," it is entirely proper to call it "Beyond Empire and Revolution." For what has taken place over the past thirty-five years is a two-stage historical transformation. The first stage was the collapse of nineteenth- and early twentieth-century empires: British, German, Japanese, Spanish, and French; and soon, the emergence of an entity, although in a still inchoate form, called the Third World. The second stage was a series of revolutions within the new nations of Africa and Asia, not to mention the old nations of Latin America, that more or less established the structure in which development would occur: within free-market or centrally planned frameworks, possessing a single-party or multiparty organization, emphasizing consumer-oriented modernization or producer-oriented industrialization—in short, all of the main issues I addressed in *Three Worlds of Development*. The purpose of this work is to examine carefully an emerging third stage, in which issues and policies arise apart from inherited consequences of empire or revolution. Hence the title is intended to be both descriptive and evocative.

Many of the issues which have captured my attention involve the words in the section titles: "Modernization, Militarization, and Mobilization." If industrialization characterizes First World systems, socialization (in the political-economic sense of reallocating wealth and power) seems with equal clarity to characterize the Second World. Militarization, in its many overt and covert forms, has underwritten a great majority of the Third World systems. In the political realm, the extent and quality of mobilizing populations and involving them in the process of development more strongly characterize Third World politics than either the electoral participation of the First World or the party-legitimation of the

Second World. Third World nations are, after all, mass societies rather than class societies, and should be so appreciated. The term "modernization" presents somewhat of a problem. "Modernization" is too close in meaning to consumer-satisfaction and transportation-communication definitions of nations to be quite serviceable. But in the larger struggle against backwardness, the word "modernization" remains appropriate. Moreover, the essence of the debate within the Third World concerns the extent to which modernization is compatible with traditionalism. That is, what are the costs as well as the benefits of modernizing a society, and to whom and for what ends? Although the dialogue within the Third World is often structured in these terms, modernization describes a very real series of processes, sometimes encouraged and at other times aborted, in relation to social change in the world as a whole.

By taking these few pages to discuss my personal excursion in search of an operative title, I hope that I have illumined the context in which I have been working, and thereby the contents of these chapters. Prefaces are poor occasions to recite what is already in a book; editorializing about one's own work is, for me at least, a particularly distasteful practice. Questions of interpretation, impact, and intent are better left to the judgment of readers and reviewers. For my part, this volume is intended to demonstrate the significant capacity of the developmental paradigm to provide a reliable framework for analysis. Perhaps it will provide answers for those who neither celebrate nor condemn the West a priori. The developmental paradigm certainly takes seriously the coming to maturity of a Third World, and is, in my view, the essential architechtonic of the twentieth century. This book is intended to elaborate and give substance to this claim.

Rutgers University I.L.H.
June 1981

Contents

Introduction: Three Worlds of Development Plus One

Serious scholarship represents both continuity and colloquy with one's own past efforts. *Beyond Empire and Revolution* is no exception. It continues a line of analysis begun with *Three Worlds of Development* (1965). One should not presume that a reader has the same sense of continuity or colloquy as a writer. Many readers will come to *Beyond Empire and Revolution* without any background in my previous work. Hence, in an effort to accommodate new readers unfamiliar with the major concepts of *Three Worlds*, I would like to present these in structural and processual terms without attempting to repeat the main lines of argument. In addition, I have elaborated upon my earlier approach by describing a fourth world dealt with only in passing in my earlier work. In the main, I should caution the reader that this morphological summary is just that; it should not be viewed as a summary of the earlier work, merely its bare bones.

Any effort at typification at a universal level is bound to create many problems. It is evident that the old nations of the Third World (Latin America) differ profoundly from the new nations of the Third World (Asia and Africa) in many respects, from religion to military organization. Such differences will be examined in subsequent chapters, when the social sub-systems are discussed. It is important to establish a cross-sectional view of how the three worlds of development are organized, before explaining their interactional patterns.

I

The First World is defined by the United States, Canada, and Japan, including allies in Western Europe.

Economy

The First World systems are capitalist, based on private ownership in a free market, development of corporate wealth, and strong tendencies toward monopolistic control of large private power blocs. The economy is typically based on individual entrepreneurial initiative, with little centralized planning (generally only in social welfare sectors). Services and consumer goods production are emphasized. The domestic market is open; a high degree of international integration is created by financial and commodity markets, with slight regulation. Saving policies are individual-voluntary and/or based on a democratic tax system. Investment is private, unrestricted, and uncoordinated. Sources of investment funds are varied, based on decisions taken by households, businesses, financial institutions such as banks, and the international capital markets. Currencies are stable. Establishing terms of trade that are mutually favorable is difficult, given protectionist ideologies, but unfavorable terms are much less inclined to have negative consequences than in Third World countries. Growth rates tend to be low, but seem adequately balanced for the needs of an industrial society. The agricultural sector is 5 to 20 percent of the GNP, and anywhere from 6 percent of the labor force in the United States to a high of 25 percent in France. The decline of the agricultural sector is driven by international competition (market forces) and by government subsidy. Industrial development is high, although industrial growth has leveled off, with the notable exception of Japan. Consumption is encouraged as a necessary stimulus for growth, but restricted by market fluctuations and private savings. Labor unions are strong, highly organized, and constitute an economic force, but only in the sense of maintaining the status quo.

Polity

Parliamentary democracy is the norm, established by legal authority and based upon economic laws of the market. Ideologically conservative and centrist, the polity accepts conflict within the rules of the game. Emphasis on the nation-state is moderate to slight, and most evident in reaction to external threat. Nations of the First World tend to have few major political parties, with strategic rather than ideological differences between them prevailing. Policy formulation is by parties and/or associational interest groups. The range of party activity is narrowly political. The party elite is small and inactive except during elections. Party organization for the most part is highly bureaucratized, and in the more vigorous portion of the First World, membership cuts across class and racial lines.

The basic party unit is the caucus. Elected representatives forming party leadership tend to come from traditional classes but remain sensitive to public desires in policy-making. Policy implementation is achieved by bureaucrats functioning under political executives. The judiciary is independent. Religious organizations do not participate in governmental or political functions (church and state are separate). The military is subordinated to government and must act as an interest group in seeking favorable political policies.

Society

The First World is highly urbanized with many medium-sized cities (250,000 to 1,000,000) in addition to several large cities. The middle class is expanding rapidly in both professional and entrepreneurial sectors; lower classes are declining as a proportion of the society. There is a high degree of vertical upward mobility and some downward mobility, increasing specialization of occupations, and some great status differentiation through occupation and income. The main avenue of mobility is through liberal mass education, in a highly bureaucratized society. Education yields high income differentials and is greatly in demand. There is an increase in leisure activities, a decrease in work activities, and considerable separation of work from private life. Social welfare programs and benefits are increasing. There is a high level of mass consumption.

Geographic mobility is high. The family unit is small and nuclear, and its emotional role is substantial. Highly organized, functionally specific independent groups proliferate. The birth rate and death rate are low; there are few officially sanctioned demographic policies. Immigration is restricted by quotas, and emigration is not restricted. Religion is predominantly Christian. Racial composition is predominantly white, and the predominant ethnic group is Northern European. Cultural affinities are Puritan and secular enlightenment.

Military

The main characteristic of the military is its professionalization; that is to say, it generally confines itself to the execution of decisions made by a non-military, civilian body. The armed forces do not have any specific ideological commitment apart from the defense of the nation defined by civilian political elites. The armed forces operate within a well-defined organizational code, with minimum emphasis on personalities, except in wartime and for propagandistic purposes. The standing peacetime army is generally small to modest in size; large-scale conscription is confined to crisis periods, although the definition of "crisis period" tends to be elongated in the present epoch. The centers of actual military power are dispersed throughout the nation, with a bureaucratic core located in the capital city. Engagement in formal, open political activities is consciously eschewed and considered a conflict of interest. Service rivalries exist; but such rivalries are normally confined to which sector of the armed forces gets what and when. There is a noticeable absence of friction between ranks, of "barracks uprisings" against higher military echelons, or open rivalries between military leaders. Requirements for admission and promotion are clearly defined and broadly accepted. Esprit de corps is very high, providing a cohesive element at critical junctures.

II

The Second World is dominated by the Soviet Union, including allies and/or satellites primarily in Eastern Europe and parts of Asia.

Economy

Second World systems are socialist, based on public (state) ownership in a strictly controlled and external market, centralization of all sectors of the economy by political elites, and total planning. The economy is typically geared to proletarian values. Heavy industrial production is emphasized. Internal and external markets are highly regulated, on the basis of economic considerations. Trade agreements are largely confined to formal allies and are not subject to severe fluctuations. Savings are non-voluntary, state-determined, and government-controlled. Investment is confined to the public sector and is highly restricted and coordinated. Sources of funds are stable, based on public budgeting. Currencies are generally stable. Imbalances resulting from the high growth rate are controlled through economic regulation. The agricultural sector is 15 to 30 percent of the GNP and 40 percent of the labor force; reductions are due to prohibition of individual private agriculture. Land speculation is discouraged in order to release resources and increase output at a rate integrated to industrial expansion. Direct and manifold controls on production and distribution of consumer goods are maintained to discourage consumption as a deterrent to industrial development. Labor unions are perceived as part of the state directorate, although worker councils play a minor role.

Polity

Democratic centralism is the norm, legitimated by rationalized authority, based upon an economically grounded ideology of communism or socialism, and interpreted by a ruling elite with doctrinal rigidity. Very strong emphasis on building common forms of economic socialism in a brotherhood of political nation-states is most prominent in times of independent political dissension of a fellow member of the Second World. Opposition is prohibited and suppressed. A strong one-party system dictates and articulates all interests. Discipline tends to be very strong, and centralization is from the top down. The basic political unit is the cell. Although working-class oriented, party elite membership is large (up to 10 percent of the population); leadership is based on bureaucratic principles of selection, and elites rise solely through party lines.

Party activity is all-encompassing and continual: policy-making is never separated from policy-implementing, powers are not separated, and legislation is from above and achieved through bureaucrats functioning under political executives. The judiciary is not independent. The army is entirely subordinate to civil control. Religion and church are subordinated or suppressed. Enthusiasm and commitment to community as well as national affairs are highly praised and broad based.

Society

The Second World is highly urbanized with large growth in medium-sized cities, and generally with several large ones. There is still a sizable peasant sector. The pace of industrial growth is rapid and is just beginning to taper off at a high level of development. The working class is growing rapidly, the middle professional class, non-bureaucratic white-collar workers, moderately. There is a high degree of social mobility—greater than that in the First World—resulting from technological changes, slowly increased specialization of occupations, and a lower degree of status differentiation. The main avenue of mobility is through the occupational hierarchy, although education is becoming more important with increased specialization, complexity, and bureaucratization of the industrial structure. Technical-scientific, mass education is emphasized. The educational sphere is unique in that it alone can claim manpower priorities over economic production, based on a notion of education as postponed economic improvements. There is a slight increase in leisure activities, little decrease in work activities, and general separation of occupational and private spheres. Social welfare benefits are high. There is a high level of mass communication and a low level of mass consumption. Geographic mobility is slight. The family unit is the slightly extended nuclear family (often resulting from limited housing). The birth rate and death rate are low; the state employs a variety of policies to integrate population increase with industrial growth rate. Immigration and emigration are prohibited. Religion is Christian; racial composition is white; and the predominant ethnic group is Slavic-Central and Eastern European. Cultural affinities are Byzantine and secular enlightenment.

Military

The main characteristic of the military is its professionalization. It is generally confined to executing rather than formulating policy, but works closely with the political elite at the decision-making level. The armed forces do not have a specific ideological commitment apart from the general social commitment to socialist ideology as defined by the civilian political elites. The armed forces operate within a well-defined organizational code, with minimum emphasis on personalities, except in wartime and for progagandistic purposes. On the other hand, the political elites often simulate military codes of dress and behavior. The standing army is generally larger than that in the First World; but mass conscription is confined to crisis periods. The centers of actual military power are dispersed throughout the nation, with a bureaucratic core located near the capital city. Engagement in political activities is considered a role of the military elites, and is eschewed among the rank-and-file (with the exception of political commissioners, either civilians attached to the military base or non-commissioned officers of political orientation within such bases). Service rivalries are frowned upon, and branches of the military are organized in terms of functional activities. The political system discourages such rivalries by insisting on military loyalty to state and party. There is a noticeable absence of friction between ranks, barracks uprisings against higher military echelons, or open rivalries among military leaderships. Requirements for admission and promotion are clearly defined and broadly accepted. Esprit de corps is very high and provides cohesion at critical junctures.

Both the First World and the Second World are exhibiting an increasing tendency toward politicization of the military. The revelations by the post-Stalin leadership of political appointments in the Soviet Union which led to disastrous bureaucratic bungling during World War II indicate a far from smooth-running military establishment. The insistence on ideological criteria for advancement to high military positions diminishes the separation of political and military elites.

III

The Third World is a group of non-aligned and non-satellite nations in Africa, Asia, and Latin America, a spectrum conventionally covering Algeria to Yugoslavia in economy and India to China in polity.

One of the most characteristic aspects of the Third World is its unwavering emphasis upon and faith in solutions provided by rapid economic growth. In this, the parallels with both the First and Second Worlds are clear: a belief in precisely those forms of activity geared to accelerate economic activity and gain. Whatever the specific international characteristics of the polity, such Third World countries as South Korea, Brazil, Nigeria, Saudi Arabia, and Indonesia have in common growth rates often twice those of the First and Second Worlds (in the area of 9 to 10 percent rather than 4 to 5 percent). To be sure, starting points are often much lower than those of the advanced nations, and the sources of wealth are frequently based on natural resources rather than industrial production. However, the Third World is, by self-definition no less than objective indicators, a significant part of the company of players which comprises the modern world economy. By extension, the social, political, and military problems they face are also similar in structure if not always in form to those of the advanced nations. Thus the essential differences of the Third World with respect to the First World and Second World take place within a context of familiarity to the "world" as we know and perceive it—a set of issues ranging from military buildup to uneven distribution of wealth will preoccupy the modern conscience.

Economy

Third World systems tend to be mixed economies, having both private and public sectors, but inclined toward some form of socialism. The choice of economy demonstrates conflict between desire for integration based on self-sufficiency, and need for aid in development. There is a tension between attempts at strong internal regulation (foreign/internal investment ratio) and moderate external regulation (freedom of block trading), and ability to implement development schemes. An additional problem in this respect is the difficulty of controlling wealthy nationals, which in turn magnifies problems in achieving better overseas arrangements. The Third World has limited choices in terms of trade and balance-of-payment problems. A variety of planning techniques exist (total, national flow, national budget, project, labor, etc.). The economy is typically based on the peasant sector; economic production is primarily agricultural, although basic industry is emphasized. The drive is to decrease the primary sector through the mechanization of agriculture, overwhelming economic push factors, and imposed economic plans, although the rate of collectivization is presently diminishing; agriculture is often neglected in favor of emphasis on basic industrial development (the highest premium is placed on heavy industry whatever the human costs) to create import substitution mechanisms. The success of the shift from agricultural to industrial sectors is not certain, and is dependent upon national planners' control; problems are compounded by the "demonstration effect," which government often tries to dampen through inflationary policies. Sources of funds are variable and generally short term. Foreign aid and investment supplement domestic savings, which are mitigated by unstable currencies. Savings policies are mostly voluntary owing to an inability to control savings. Growth rates are high in percentage terms but not in per capita yields. Attempts are made to control imbalances through political regulation. Short-run private-sector gains often vitiate effects of long-run public-sector planning. Labor unions are radical and unstable, and play a small economic role.

Polity

Mass democracy is the norm, legitimated through highly articulated and politically grounded ideologies of nationalism and socialism, and focused on a charismatic leader. Ideology is radical and socialist, but with considerable variation; it is polycentric rather than monolithic. The basic political unit is the party branch. The state is highly centralized and under virtual one-party rule. The elite desires total control and suppresses conflict groups, and the lack of disjunction between policy-makers and policy-implementers is mitigated by a number of factors: low level of party discipline; enthusiasm and commitment, often attached to party personages rather than to party policies; absence of institutionalized charisma; and responsiveness to the desires of the masses. Change in policies is due less to minority parties than to articulation and application of multiple interests. The military, executive branch, and bureaucracy are often independent and over-participant. Although oriented toward the "popular" class or the nation as such, party membership ranges between low in the First World and somewhat higher in the Second World. Leadership is often based on educational-technological-military criteria. Except under highly revolutionary circumstances, the leadership makes alliances with elites of traditional prestige unless the latter are resistant to development.

Society

The Third World is rapidly urbanizing, with the problem of developing middle-sized cities. Often the major city is more powerful than the nation as a whole. Industrial growth is still approaching the "take-off" stage. Social mobility is still low in traditional areas; occupational mobility is slowly increasing in industrial areas. The labor market remains undifferentiated. Traditional status distinctions are diminishing, and the main avenue of mobility is occupational and political hierarchy.

Emphasis is on mass literacy and technical-scientific secondary education, with planning to integrate industrial needs into educational programs. Leisure activities are few; work activities are increasing, often with no clear distinction between occupational and private spheres of life. High social welfare benefits characterize Third World nations. Mass communication is at the developmental stage. The extended family predominates but is deteriorating. Birth rates are high and death rates are declining (although severe problems of mortality still exist in major Asian countries like India, Pakistan, and Burma). Third World nations exhibit great difficulties in establishing birth-control policies. Immigration and emigration are generally restricted by occupational skills. Religion is Moslem, Hindu, Buddhist, and Christian (in Latin America). Racial composition is yellow, black, brown, red—in short, basically non-white. The predominant ethnic groups are Negroes, Arabs, and

Indians. Cultural affinities are philosophical rationalism closely linked to religious expressions. No clear separation exists between the secular and the sacred as in the Western tradition, but there are fusions between theological doctrines and political credos.

Military

The military is politicized, that is, it is a formulator and executor of political decisions. "New nations" witness the rise of military politics, either as a prime mark of sovereignty or directly, as the ruling political group. Armed forces have specific ideological commitments—in pre-revolutionary countries to conservative Caesaristic models, in post-revolutionary Third World countries to radical-charismatic models. The organizational code of the military is ambiguously defined, with an extreme emphasis on personalities, on the military leader as national redeemer. The standing army is relatively small, but with a disproportionate ratio of elite to rank-and-file. The level of financial expenditure is quite high considering the size of the standing army—in part owing to the need to purchase arms from foreign industrial powers, and in part as a consequence of maintaining an agency capable of maximum coercion with minimum numbers. The centers of military operations are usually concentrated in or near the capital cities or industrial centers, enhancing the possibility of "palace revolutions." Such concentration is also a symbolic reflection of military power near the centers of political decision-making. Political activity is most noticeable at the elite level, but often in periods of discontent, rank-and-file leadership will manifest itself. Service rivalry is exceptionally high, with the navy and air force often taking positions to the political Right of the army. Friction among military elites is considerable. Factionalism oftentimes becomes noticeable before a coup d'état. Esprit de corps is not particularly strong—except within "in-group" or barracks factions.

IV

The Fourth World is not dominated by any particular nation or cluster of nations since its unifying characteristic, if one can call it that, is lack of dominion and often lack of sovereignty as such. It includes approximately fifty to sixty nations throughout the world, forming a tropical and sub-tropical belt around the equatorial portions of the world, extending from the Amazon region to the Sahel in Africa to Kampuchea in Southeast Asia, sharing basic human needs and wants. Their main characteristic is that these areas produce insufficient energy and inadequate food, and hence are basically recipient rather than donor nations in every fundamental sphere.

The most distinctive properties of the Fourth World are its place in the universe as a recipient rather than donor nation; its incapacity to generate a surplus of funds which can then be negotiated for other basic needs; and its emphasis on group survival problems rather than national sovereignty problems. There are all-too-familiar extremely low levels of formal education, surface and air communications, and medical and health care, which results in extremely high rates of malnutrition, disease, and infant mortality. The Fourth World is characterized by indicators at the opposite end from those of the Three Worlds in development; perhaps worse, a seeming inability to break through such barriers, or even to recognize benchmarks of development as such. What the Fourth World reveals is not a continuum with the remainder of the world of development, but a sharp break with that world; both in material terms which are easily measurable and complex spiritual terms which rationalize modern life, such as the place of scientific discovery and innovation in the affairs of peoples and nations within the three worlds of development. If the Third World can be called a triumph of the present century, then the Fourth World must be viewed as its essential tragedy.

Economy

Much of the Fourth World can be called pre-industrial or at least outside the area of capitalist and socialist competition characteristic of the other three worlds. Land for the most part remains under tribal and/or private ownership and political/economic proprietary rights are largely undifferentiated. For the most part production is agrarian, characterized by hunting and gathering techniques rather than scientific land management or crop rotation. Almost all goods produced and services performed are absorbed internally and locally. There is a low level of national integration and almost none with other nations. Fourth World economies are generally characterized by an absence of overall national policy and a low differentiation of social and political strata. Investment is unusual; the level of the economy does not permit savings, taxation, or the kinds of managerial coordination that permit business and banking. Many Fourth World economies are characterized by low gross national product; oftentimes it is at negative levels. At the same time, high population growth, without coordination with the economy as such, makes economic growth exceedingly difficult. The orientation is toward production rather than consumption, based upon essential needs rather than consumer wants.

Polity

Fourth World societies tend to be based on traditional modes of authority, with informal political organizations matched by religious, ethnic, or tribal associations managing the affairs of community and individuals. The political system is sometimes characterized by a rough democracy that comes about through the absence of centralized authority rather than institutionalized equity. Centralized authority is found only in a small urban sector without much power over other sectors.

Society

Fourth World societies tend to be rural, labor-intensive, with low levels of urbanization and modernization. The major social problems are caused by the need to purchase both finished commodity goods, industrial goods of all kinds, and the energy to produce such goods, which leads to a never-ending cycle of repressed wants. The main strength of these societies derives from having rare energy resources. Even if there is no monopoly for such resources, they provide basic income and return on investment. Societies are characterized by particularism and localism in religion and in cultural patterns. These show remarkable similarities with primitive cultures elsewhere, but are largely delineated by local factors. Racially, most of these societies are non-white.

Military

Fourth World armed forces tend to be weak, but have properties similar to the military of the Third World. They have a high degree of politicization, characterized by the concept of the military as police, essentially charged with internal maintenance of order and taxation rather than a specific ideological commitment to overseas activity or even national development from above. Unlike the Third World, there is little sense of the armed forces as part of an ideological commitment to global ambitions. But like the Third World, the military tends to be relatively small, with a high ratio of an elite officer-corps to rank-and-file membership. Oftentimes Fourth World armies are satrapies of Third World military regimes.

The stability of the Fourth World is mantained by several factors: (1) the enormous costs involved in managing these societies; (2) the reticence of larger nations to become directly involved; and (3) the interplay of socioeconomic forces within the Third World which come to view Fourth World sovereignty as protected by a variety of forces within the Third World. These powerful entities within the Third World attempt to maintain stability by excluding Fourth World nations in major decision-making on a worldwide scale.

The Fourth World, by virtue of its increasing involvement in the world system, no less than its energy and food shortages, contains the most serious problems within the global system. It can no longer be viewed simply as an extension of the Third World, which has itself become heavily fractured into both donor and recipient nations in the international economy. For example, OPEC nations from the Middle East are donors within the international economy, while high food-producing nations in South America are equally involved in the international economy as donors (as well as recipients). It would therefore be a conceptual and practical error to consider the Fourth World simply as a special variant of the Third World. The Fourth World represents a dimension of powerlessness and economic attrition that, while unimportant in the play of economic forces, is exceptionally important in the determination and delineation of political forces. Hence it is very much part of a global system of power even if it may not exert an independent dynamic at international levels.

This, in brief, summarizes the characteristics of the four worlds constantly alluded to in *Beyond Empire and Revolution*. This rough typology must necessarily be elaborated and re-examined on a continual basis; we are not dealing with statics, but with dynamics. Countries like China move between Second and Third World dimensions, Poland moves between Second and First World dimensions, India moves between First and Third World dimensions. Analyzing these shifts requires sensitivity to national varieties, cultural forms, and patterns of evolution that must sometimes be addressed regionally, sometimes globally. Defining the major blocs within the world provides a shorthand language for discussion and analysis of major issues. This shorthand cannot, of course, be substituted for the concrete empirical research required to make the character of international society, economy, and polity come alive and vibrant.

I

Modernization and the Limits of Development

1

Realignments of Three Worlds in Development

Rich and poor nations are now defined very differently from the way they were in the early seventies. The underdeveloped portions of the Third World lack sufficient energy or foodstuffs for export purposes, and commonly lack supplies adequate to support internal growth. So-called poor members of the Third World make three main demands: (1) a common fund to pay for ample stockpiles of necessary commodities, which has been called "commodity stabilization"; (2) a moratorium of indeterminate length on the debt they presumably owe to rich countries; (3) "indexing," a linkage in direct fashion of commodity prices to overall levels of inflation. The non-developing portions of the Third World seek to impose these three demands on the developed world. Meanwhile, previous Third World solidarity, while always unstable, has completely given way to splinter and separation, making it even harder to impose demands on the industrial West or on the socialist East in any unified way.

A common fund to pay for stockpiling commodities would be of some advantage to the First and Second, as well as the Third World of development, in theory if not in practice. There are, however, practical limits to the advantages. Such a common fund could bring stability to coffee, rubber, cocoa, and copper markets and, by smoothing out price fluctuations, guarantee a flow of profitability to producing countries and a flow of goods to consuming countries. Exchanges of raw materials and finished goods could take place at mutually agreed upon levels of profitability, but currently the principle of commodity stabilization remains vague and ambiguous. Smoothing out fluctuations in the price of coffee might be simple enough, but OPEC nations would object to the application of such a principle to oil. OPEC has shown it has difficulty agreeing to price structures among its members; the prospects for OPEC's rationalizing oil production to the advantage of the West or the energy-poor regions of Asia and Africa are dismal. The Fourth World has come to appreciate that the OPEC model may not be replicable. The control of a large portion of oil supplies by a few nations is real enough. Other crops or minerals are not so easily monopolized, being relatively easy to produce or transfer from one area to another and having ready substitutes in many cases. It is much more difficult, although technologically possible, to substitute solar energy for crude oil. Efforts to achieve a common world price for basic commodities or stabilize prices in a floating market must confront the fact of profound differences in the allocation and availability of resources within the Third World, which make the procedures well-nigh impossible. Industrial nations may seek and get safeguards for further investment in the developing world as well as assurances from oil-exporting countries about future price stability for petroleum, but in negotiating with the disenfranchised and non-developing sectors of the Third World they are at an impasse.

A debt moratorium is also easier to set forth as a goal than to implement as a policy. Wealthy Third World nations such as Mexico and Brazil oppose a debt moratorium because it would weaken their short-term credit standing in the international financial markets. It would be far simpler for the First World to absorb a debt moratorium than for the Third World to accept it. The debt would have to be repaid eventually, and in the short term a moratorium on debt would enlarge the scope of First World control, even making domination of parts of the Third World possible, since the loans would continue to accumulate interest. Only repayment of the principal would be held in abeyance. Rapidly developing areas of the Third World need to establish a firm line of credit. Long-range political considerations must yield to short-range economic requirements. The dilemmas and contradictions within the Third World make international economic cooperation difficult to implement.

Indexing, or linking commodities to rates of inflation, would also have problems. Such a linkage would accelerate inflationary tendencies. It would create spiraling costs and prices rather than mechanisms to prevent inflation. Beyond that, indexing would transfer real wealth from the First World to the Third World, without discernible benefits to the former. It would have the paradoxial effect of depleting Second World credit balances by forcing nations to enter the commodity market to purchase goods at higher prices because indexing is pegged to hard Western currencies. Second World interests would not be served by an indexing procedure that would curb inflation and rationalize the First and Third Worlds at the expense of the Second. The Second World would oppose an indexing process. Little wonder that at international gatherings, delegates from the Soviet bloc argue that the plight of poor countries is entirely due to economic exploitation, and that it is up to capitalist countries to make good on their promises. Implicit in their criticism is an assumption that the socialist bloc has no role to play in the development of underdeveloped countries, which is a far more comfortable posture than admitting that the Soviet Union is least advantaged by the sort of impulses which inspired the series of conferences on international economic cooperation. To take a contrary position would imply that the Second World is aligned with the advanced industrial powers as a donor nation. Furthermore, the Second World would bear the negative consequences that might stem from such a redistribution of international wealth. It is not surprising that the Second World denies it has a role in the development process.

Certain fundamental realignments leading to major structural changes are in the making. Wassily Leontief has estimated that by the year 2000 the developing world will have doubled its Gross Development Product; similarly, most of the developed world economies will have expanded, but at much slower rates. Hence the developed world's share of the GDP will then be reduced from its present proportion of two-thirds of world levels to one-half.[1] From a Third World standpoint, this is an optimal scenario. But even when we divide the Third World into a variety of developing markets, developing areas of the Third World such as the Middle East and portions of Latin America on the one hand and the arid and tropical sectors of Africa on the other, we see that they have different goals. The statistical trends underscore powerful ideological differences between successful and unsuccessful sectors of the Third World. So clear-cut is this schism that the Third World of development is quite distinct from what may be called a Fourth World of non-development.

There are serious problems with commodity strategies. Among the very serious is the pot-of-gold mentality inspired by OPEC and oil-producing nations. Thus far the oil situation is unique in that oil is not replenishable and it is not easily replaced. Rubber, for example, is easily replaced by nylon and rayon substitutes; sugar by saccharin, coffee by tea and cocoa, steel by aluminum. The problem for those with pot-of-gold expectations is that oil can be replaced by solar and electric power, and even worse, petroleum itself is in short supply. The most optimistic engineering reports indicate there will be a substantial shortage in traditional energy supplies by the turn of the century.

The acquisition of financial capital, without going through the process of industrialization, led to a dramatic shift in Gross National Product, but no corresponding distribution of wealth. The Third World became restructured as a result of financial shifts of great magnitude. Asymmetrical relationships within Third World nations became stretched to breaking point, leading to either revolution or repression. What took place, and here nations such as Saudi Arabia and Iran are typical, was that the excess wealth created by oil surpluses was spent in the First World to purchase industrial equipment and manufactured goods. The wealth was used to pay for technology transfer and corresponding industry skills which drive up the prices of finished goods. OPEC-inspired wealth creates new forms of exchanges of goods and services, but not necessarily profound changes in the social system.[2]

The *nouveaux riches* of the Third World have been plugged into the *anciens riches* of the First World. The politics of mineral wealth in general have enlarged world capitalist markets and accelerated an exchange of goods and commodities among elite sectors within the First and Third Worlds. While problems of growth are thereby partially solved, the problem of equity remains largely unresolved. The sophisticated strategy of the First World has been based on the realization that new Third World wealth is not automatically a disaster for the First World. This strategy has been to channel discussions of complex Third World demands into existing multilateral institutions such as the World Bank and the International Monetary Fund. These institutions become repositories of economic rationality, clearinghouses for the exchange of goods and services. While this approach accommodates wealthy nations of the Third World such as Iran, Brazil, or Saudi Arâbia, it only enlarges the gap between growth and equity demands. If it does not divide the Third World into a Third and a Fourth World, it at least points up profound differences between wealthy and poor nations within the Third World. The World Bank and the International Monetary Fund want balanced budgets and tight constraints on social welfare spending. They want what any conservative fiscal manager requires for further investment, with the result that contradictions within the Third World become graver. For example, the United States position under the Carter administration was to negotiate only those loans cleared through the International Monetary Fund. As a result, the IMF became a credit clearinghouse for loan eligibility. This imposes certain constraints that militate against the Third World as a unified bargaining agent, and most certainly as a unifying political agent.

Within such a context, tensions between the First and Third World—between nations at opposite ends of the energy and food resources spectrum—become acute and lead to the possibility of a new series of alliances based on material wealth rather than ideology. This creates special problems for the Second World. The Soviet Union can neither afford nor create commodity relationships that produce high-level surplus. Even more than Eastern Europe, the Soviet Union is in the uncomfortable position of describing itself, in the words of Leonid Brezhnev, as a developed socialist society, when in fact it is isomorphic with underdeveloped Third World societies with respect to levels of commodity goods production. Even Brazil is outstripping the Soviet Union in quality and quantity of goods and services.

The Soviet Union's inability to establish pre-eminence either in industrial manufactured goods or in basic foodstuffs for export has led to a situation in which production within the COMECON

(Warsaw Pact) nations is too low in quantity to meet domestic demands and too poor in quality to be exported. The flow of goods and services from the First World to the Second World during the second development decade (the 1970s) was more impressive than the flow from the Second World to the Third World. Multinationalism dictated COMECON policy; everything from soft drinks to automobiles was imported. From an economic point of view, the socialist bloc is shrinking. Although it may be registering certain political gains and aims, it has not been able to create a base of surplus wealth that can be used as a bargaining chip, to entice the Third World in a Second World direction economically, or to offset the possibilities and capabilities of the First World to move the Third World in its direction. The comparison of OECD with COMECON production figures indicates the serious economic lag of the socialist world. For the first time, the First World is beginning to exploit the political consequences of that lag.

In an unusually candid speech, Egon Bahr, father of West Germany's Eastern policy, lectured seventy members of the Soviet Academy and its Institute for the Study of the United States and Canada on the facts and figures of Soviet failure to address the problems of the Third World. West Germany gave two and one-half times as much for development aid in 1975 as did all the countries in COMECON. Between 1974 and 1975 the First World increased its contribution to overseas development from $11.3 billion to $13.6 billion, whereas COMECON nations dropped theirs from $1 billion to $0.8 billion. The Soviet Union has been especially delinquent. According to European Common Market data, Soviet net payment for development aid declined from $650 million in 1972 to $350 million in 1975. At the 1977 council meetings of the United Nations Commission on Trade and Aid, the bloc of seventy-seven developing nations informed the socialist countries that they should recognize a new era in which all countries had to be concerned with Third World problems. Yet the Soviet Union has expanded only its ability to supply arms to the Third World. Of the Soviet GNP, only 0.05 percent is earmarked for development aid, whereas First World nations give developing countries at least 0.3 percent of their GNP.[3] As the rhetoric of imperialism and colonialism rapidly gives way to a hard look and stare at *which* nations are providing *what* amounts of aid, the Soviet Union's lack of participation in development aid must be seen for what it is: a monumental failure covered by a multitude of words.

The inability of the Soviet Union to join with industrial nations to solve the problems of the Third World will ultimately mean that the relationship between the First and Third Worlds will become closer and stronger, especially for those members of the Third World able to compete on approximately equal terms with advanced nations. In fact such nations often participate in a free market world economy on a par with the industrialized West.

The first development decade (the 1960s) represents a series of strategies converted into an ideology. It involved demands for extensive aid. People like Raúl Prebisch articulated a moral demand by the Third World upon the First World for aid of all sorts—monetary, commodity, foods, etc.—and the veiled threat that failure to come up with that aid would lead to socialist revolution. This demand established firm ties of dependency between donor nation and recipient nations. There was no rush toward socialist economic choices, although there was considerable movement toward socialist political organization. This was the legacy of the first development decade.

The second development decade witnessed forceful demands for equality in trade. They sprang from a strong set of Third World beliefs, sometimes correct, other times not, that there should be greater parity between basic foodstuffs and mineral resources on the one hand and finished industrial goods on the other. The second development decade proved far more efficacious and workable precisely because it established a *quid pro quo* rather than simply relying on a moral imperative.

While development of sectors in the Third World capable of supplying finished commodities and raw materials alike accelerated, these Third World nations also became involved to a much greater extent in the world capitalist system than they had bargained for. The second development decade, while manifesting a vast improvement and a considerable step forward from the first development decade, nonetheless had its own problems. Not only did demands escalate at a quantitative level, but the international distribution of wealth affected its character. At the same time the West is moving toward greater awareness of the value of energy and foodstuffs and of the ecological setting in which agrarian production takes place, the Third World is reaching out for a greater share of commodity production and industrial productivity. The development strategy for the 1980s will involve increasing pressure for the redistribution of profits, goods, and services, and by the end of the decade, there will be increasing demands to shift industrial production to a large chunk of the Third World. Demands for redistribution of profits will mean a change in economic arrangements, so that the Third World will not simply be a supplier of commodities and the First World a supplier of industrial goods. However large in importance the agrarian sector has loomed in recent years, industrial growth remains the bedrock of genuinely wealthy nations. The lesson the socialists of the Second World imbided from the First World has not been lost on the Third World.

Within the larger framework of the distribution and production processes, and as a forerunner to demands for a third development decade, the energy, ecology, and environment crisis has already led to an international redistribution of wealth. For the first time, although probably not the last, such a massive redistribution has occurred without a conventional world war. World War I made American capitalism supreme; World War II provided the Soviet socialist societies with a major geopolitical role in world affairs. "World War III" was fought with boycotts and embargoes, although not incidentally by guerrillas against organized political systems. It gave a section of the Third World autonomy and an area-wide integrity in the large land masses of Asia, Africa, and Latin America. This considerable victory of the Third World enabled it to redistribute international wealth without actually having fought conventional battles. These changes were achieved at the price of deferring equity demands within the new nations. With no sense of irony, the World Bank, the presumed apotheosis of imperial policy, stepped forth to enunciate a position implying that investment decisions would henceforth be based on equity and not just growth. Its president, Robert S. McNamara, issued a statement in June 1977 that "while economic growth is a necessary condition of development in a modern society, it is not in itself a sufficient condition. The reason is clear: economic growth cannot change the lives of the mass of people unless it reaches the mass of people." A major economist and vice-president for development of the World

Bank, Hollis Chenery, remarked in clarifying McNamara's position: "We are looking at what governments are doing on income distribution, and it is one of the considerations now in making project loans."[4] The charge of "outside interference" has not yet been heard from those ideological sources who automatically place responsibility for exploitation on the shoulders of such agencies as the World Bank. This leads one to suspect that conventional ideological persuasions do not explain present problems in the development process. On the contrary, equity considerations are now central to United States decision making on all foreign aid projects.

Implementation of this policy is most threatened by forces in the private sector that have benefited from the current arrangement. The growing influence of private transactions in the international monetary market makes it practically impossible to exercise control over international liquidity, however desirable that end might be. As Carlos Massad of the Economic Commission for Latin America has noted: In 1964, private holdings of international liquidity were one-third of the total official reserves, whereas a decade later, in 1973, private holdings amounted to two-thirds, and far exceeded official reserves of the nations in whose currencies such holdings were maintained.[5] This leads to a transfer of resources from developing countries to those whose currency is being accumulated to attain fiscal stability. As a result, the world economic system tends to underwrite current emphasis on growth at the direct expense of equity considerations.

The present division within the Third World leads to a sharp differentiation between have and have-not nations at the level of industrial production and wealth. It exposes an unwillingness by the Third World to help a Fourth World that is not developing. The nature of alliances and alignments must shift, in the short run at least, to the benefit of the First World, and indeed it already benefits from certain contradictions that exist today. The Second World is unable to generate sufficient exports and the kind of wealth that would attract new societies that have wealth—the Middle East, for example, or Latin America.

The First World must learn how to live with a shrinking portion of the international financial pie. The more aid that is supplied to the Third World and the more autonomy achieved by the Third World in terms of production, the more difficult the position of the First World with respect to its own internal patterns of consumption. As consumer demands of the industrialized West grow, the ability to satisfy them shrinks. That is an irreducible, ineluctable lesson of the early 1970s, and certainly the main lesson of the oil boycott.

The present game of international redistribution of wealth includes new players—players unwilling to include most of the disenfranchised. As a result there exists real but quite limited progress. The 1980s will be marked by severe contradictions within each world of development: so much so that they may actually break up and bring about new realignments and new blocs. Within the capitalist world, contradictions between the United States and Western Europe will become even sharper. Every level of production, whether it be supersonic jets, computer hardware systems, or chemical fertilizer plants, will be marked by growing competition of the United States with Western Europe and Japan as well as with the sectors of the rest of the world.

Tensions within the socialist orbit, specifically between the Soviet Union and the People's Republic of China and their allies, have not only not abated, but they have actually been intensified by transfers of leadership in both nations. Again, at the economic level, there is competition between a relatively wealthy nation such as the Soviet Union and a relatively impoverished one such as China. At the same time, problems of affluence itself have made the Soviet political system less workable and less capable of high mobilization and high output, whereas the Chinese with their capacity for high mobilization cannot convert success in mobilization into real wealth for large numbers. Yet more successfully than the Soviets, the Chinese have created equity among all sectors of Chinese society, and thus achieved a major goal of socialism—"from each according to his ability, to each according to his needs." The Soviet Union and the COMECON nations have moved toward a much more highly stratified society, with all the corruption presumably characteristic of the capitalist world multiplied by virtue of the ever-expanding nature of bureaucratic power in the Soviet Union.

The Third World is characterized by military systems of rule. But in turn these have to contend with an energy-rich Middle East, a food-rich Latin America, an energy-poor Asia, and a food-poor Africa. There is a peculiar realignment in which regionalism is more important in the definition of world affairs than are holistic concepts like *Third World*. Internationalization or rationalization of the world economy has not occurred. It may be dangerously optimistic to speak of a unified world economic order under the conditions of increasing scarcity of goods and services. What has occurred is increasing stratification at the expense of equity. The globalization of nationalisms is displacing internationalization of the economy. This increasing self-interest sometimes expands beyond the nation into the region. Oil-exporting nations are reluctant to continue holding special talks with consumers at which both industrial and developing nations criticize pricing policy.

The club of wealth is expanding more rapidly than the Club of Rome. This is a new round in development, and there are new players involved, from Australia to Brazil to South Africa to China. But we are faced with a different global environment, producing a redistribution of wealth. A mistake is to presume that wealth redistribution is necessarily democratizing. Nations do not improve the quality of life for their people simply by virtue of becoming wealthier; the quality of life may even deteriorate.[6] There are two meanings to democracy: one, equity within a nation; and two, equity among the players in the world scene. In the latter sense, greater democratization has been achieved. In the former sense, it has not. We have seen an enormous transference of wealth among nations in the latter part of the 1970s, but without the achievement of much real equity within nations.

2

Traditionalism, Modernization, and Industrialization

How are the terms traditionalism, modernization, and industrialization employed in Third World studies? Researchers use these terms in very different ways. Even the phrase "neo-traditionalism," or the new-tradition orientation, involves two words more often juxtaposed than synthesized. Clearly, taxonomy overlaps ideology, and vice versa. How this happens is the subject of this chapter.

The Concept of Traditionalism

The first and most obvious indicator of traditionalism is backwardness. In the work of Daniel Lerner and of Wilber Schramm,[1] the notion of tradition was correlated with backwardness, characterized by slow communication and parochialism measured by the limited radius of life traversed by ordinary people. Thus, one characteristic of tradition concerns informational and ecological backwardness.

A second indicator of traditionalism, and one that is more ambiguous, is the maintenance of devout religious belief, affecting daily life in the broadest possible way. Religious tradition per se does not have much to do with backwardness; the key factor is devout belief related to worldview, what might be called mass ideology. As Peter Berger well understands, religious belief often serves as the ideology of the masses.[2] Whether traditionalism is a better or worse opiate of the masses than secular doctrines, such belief need not preclude the developmental processes. Whatever its metaphysical aspects, religion does represent a general organizing premise, a broad sensitizing agent in everyday life. In social terms, religious tradition defines how one relates to others and to oneself, and, beyond that, how one positions oneself in the world of institutions as a whole. This notion of religion is at considerable variance with the notion of backwardness as an economic or cultural issue pure and simple.

A third indicator for traditionalism is partially entwined with the previous indicators, but is distinctive enough to merit separate identification. In his early writings Fernando Henrique Cardoso equated traditionalism with a sense of cultural identification,[3] at which level traditionalism has to do with inherited secular values rather than transcendental values. Cultural nationalism comes to represent an ongoing commitment to a long-held worldview by masses of people. In the modern era, ideology may become disengaged from theology, a notion of traditionalism extremely powerful and compelling, particularly in Latin America. Folklorism, nationalism, and cultural distinctiveness form a coherent entity quite apart from a religious belief and are linked to patterns of land tenure, language clusters, and a range of psychological attitudes about sex, race, and ethnicity.

A second major motif of Third World studies is modernization. In its strict, non-ideological sense modernization is essentially a technological term, not a judgmental one. It concerns the displacement of human labor by machine labor. Modernization is linked to the instant communication of information, the rapid movement of people and goods, the automation of services, etc. Unfortunately, the concept has been abused to imply some sort of moral revolution, an alleged superiority of those who "become modern" over those who "remain traditional." Curiously enough, the same sort of juxtaposition took place in the nineteenth century between science and religion, in which the former was viewed as a displacement mechanism for the latter. Science was declared to be "at war" with religion and theology.

The Concept of Modernization

What is meant by this concept, and to what use is it put by researchers? Each social science discipline gives the concept a different emphasis. Economists see modernization primarily in terms of human application of technology to the control of nature's resources. Political scientists have emphasized the disruptive aspects of nation building. Psychologists have emphasized the growth of knowledge, education, and individual achievement.[4] Modernization in these various guises deals with the human transformation of culture and rapid mobilization. Invariably proponents of these social science orientations hold modernization to be a precondition to development.

There is another notion of modernization which derives more from geographical roots than social science training. It is probably more commonly held in Latin America than in the Middle East. Analysts from the Middle East have tended to think of modernization as a communication-transportation precondition to development; whereas analysts in Latin America, when they refer to modernization, more often than not are talking about consumer abundance, with middle-sector domination as a precondition to development.[5] The literature on Latin America discusses modernization as invariably related to consumption, from electrification to the acquisition of material objects. The coming together of the middle sectors is said to make modernization possible. Class stratification, rather than cultural transmission, represents the essence of modernization.

A further definition of modernization, the one most apt to be advanced most in the more developed sectors of the Third World, is what might be called modernization as the tradition of the new, to paraphrase Harold Rosenberg's literary concept. Here modernization refers to urbanization, or the extent and rate at which a society is de-ruralized. Throughout the Third World have emerged cities comparable to national centers elsewhere. Improvements in health, welfare, and educational life chances are dependent on high concentrations of people and resources—in a word, they are dependent on urbanization. Modernization at this level has to do with physical locale and geographical placement of people, rather than transportation-communication, or consumerism-abundance. Modernization represents the transition from rural *Gemeinschaft* living to urban *Gesellschaft* living.[6]

As an organizational concept, modernization, like traditionalism, clearly leaves much to be desired, not because social scientists lack skills or intellect, but because the social world is complex. We have not yet figured out how to dissect a major term such as modernization qualitatively, much less quantitatively. To give it operational meaning is a further task.

The Concept of Industrialization

A third term with significant relevance in studies of the Third World is industrialism, or industrialization. My own work has contrasted the concept of modernization with the concept of industrialization.[7] I took the Latin American definition of modernization as consumer abundance as the crucial pivot, with the Soviet notion of modernization as heavy industrialization as its antithesis. In this context, the degree of modernization became *ipso facto* the extent to which a society was able to industrialize, and vice versa.

Industrialization carries with it a great many political subtleties. It shifts emphasis from the development of economic wealth to the development of political power and, sometimes, social order. For example, industrialization in south India has little to do with heavy industry. There industrialism means the concentration of a large number of small-scale entrepreneurs operating in what is considered by them to be industrialized environments which are, in fact, cottage workshops.[8] The Indians employ a kind of industrialization designed to develop a self-sustaining consumer-oriented economy, but they are continually frustrated by an inordinately high rate of population growth. Whether the object is the manufacture of clothing, pots and pans, or even radio and television, the crucial factor is local ownership and operation, without the infusion of multinational superstructures. The attempt to achieve a modest level of growth without mortgaging one's society to external economic forces runs up against the need to feed the masses and satisfy the consumer demands of the elites.

A second type of industrialization is found in the socialist world, principally the Stalinist world. This is industrialization as heavy production, that form of factory management and productivity that creates a large volume of basic industrial products and tools and in turn makes consumer goods a possible option. This type of industrialization carries with it the clear, if sometimes ambivalent, premise that consumer needs will be postponed until industrialization is complete.[9] A state does not decide to industrialize for the sake of consumer gratification, or to create a product line already produced by others; rather it industrializes as a consequence of the internal dynamic of that society. This type of industrialization thus becomes a surrogate term for modernization based on industrial output rather than on consumer satisfaction. Often such industrial growth involves a trade-off of high export sales for low internal consumption.

A third meaning of industrialization has increasingly come into play: industrialization as symbolic politics, as a statement of power, in which a society achieves an international rank order of power, depending on degree of industrialization, as industrial technology becomes a factor in military capability. This type of industrialization should be distinguished from the kind that went on in late nineteenth-century England and the United States, or in the early twentieth-century Soviet Union. This heavy industrialization has a direct component—the notion of aggrandizement or anticipatory aggrandizement of the forms of production. When the Nazi industrial machine accelerated growth in an environment that was already highly industrialized, its technological transformation was from civilian to military forms of manufacture.[10] This is a latter-day species of industrialism. Economists like Kenneth Boulding and John Kenneth Galbraith have persuasively argued that war industries are not easily convertible into civilian production.[11] During the period 1939 to 1979, aircraft industries never converted into civilian forms of activity, in part because they have generic structural components that constrain them from being converted into civil-sector forms of industrialism.

That is why in the second half of the twentieth century one can speak of a third kind of industrialization. Military industrialization increasingly characterizes the more advanced sectors of the Third World. Like consumer-oriented industrialization, it is a direct demonstration of power. Whichever type of industrialization is achieved by a nation, it is a direct indicator of political power.

Traditionalism, Modernization, and Industrialization as Ideological Choices

My purpose in the preceding discussion was to avoid using a methodological ploy that would make the world any more complicated than it is. To do otherwise violates the principle of scientific parsimony, not to mention common sense. I want to suggest how traditionalism, modernization, and industrialism can be linked to ideological statements about the uses of power. Policy making is an organizing premise for creating economic strategies to satisfy widespread demands and critical needs. Development involves problems of choice instead of convention. This is not to deny that different historical epochs lend themselves to different decisions, but rather to assert that development presents political leaders with the need to make serious strategic decisions. Traditionalism, modernization, and industrialization are susceptible to different mixes and timing. Leaders can decide (1) to satisfy certain sectors of the population rather than others; (2) to make certain international coalitions rather than others; and (3) to introduce and infuse into a national society certain extrinsic elements, such as multinationalism, or exclude those foreign elements at the risk of lower living standards and/or consumer satisfaction.

Every such decision creates structural tension and strain in the social order. No single policy will make the world a better place to live without any negative side effects. The tensions and strains involved in decision making are the everyday problems of the policy apparatus. Traditionalism and modernization at the policy-making level of political reality involve a good deal more than blind historical antecedents or normative cultural preferences; they involve a choice of systems and actions.

When practical decisions are required, and policy makers are faced with inherited paradigms (insofar as they are available), they always manage to find new combinations. The political world invariably, if inadvertently, conspires to outmaneuver theoreticians. For one hundred years, economics operated within a broad-based concept of affluence. Both neo-Keynesian and neo-Marxist concepts presume that economic problems are basically allocational in character. A synthetic developmental figure like Galbraith[12] for many years argued that the world had solved certain fundamental issues at the level of technostructure; and what remained were allocational decisions, what the nineteenth-century economists referred to as a society's "stock of know-how." These allocational decisions permit outcomes that allow the social system to survive.[13]

As a developmental paradigm crystallizes, new factors come into play, often without warning. New ideas about energy, environment, and technology profoundly and directly affect the way developmentalists think about traditionalism, modernization, and industrialization; terms seemingly analyzed beyond boredom. Yugoslavia and Hungary formerly emphasized industrialization because that is what Stalinism required; they have begun to appreciate that their real commodity value, their basic export goods, are largely in the agrarian sector. Middle East nations have begun to view their oil supplies not as a perennial miracle but as a temporary monopoly. On the other hand, Brazil, Japan, and Korea have begun to understand that cheap labor without a technology appropriate to high-priced energy will no longer make them unrivaled world economic powers.[14] Neo-Malthusianism haunts both Marxists and Keynesians.

At the same time, the issue of consumer satisfaction has forced itself on the historical agenda. Gratification can be postponed only as long as a nation believes it is surrounded by enemies. When the enemy vanishes, as the result of ennui rather than actual defeat, it becomes increasingly difficult to postpone consumer satisfaction. One can continue perhaps for one or even two generations to say that such satisfaction must be put off because "we are surrounded by enemies of our state or our system." But such rhetoric, like any millenarian claim about the end of the world, loses its conviction at some point in time. Material gratification becomes the necessary basis of rewards for work done.

Consumerism has been the curse of the socialist system since 1949 and 1959 and has produced its heresies. Those dates demarcate the second big socialist revolution (China) and the first presumably socialist revolution in the Western Hemisphere (Cuba) respectively. Both nations tried to face down the problem by creating "moral equivalents" to consumerism. In fact, both societies temporarily elicited what Marx would have called unpaid labor time.[15] That is to say, political leadership translated moral fervor into a denial of mass consumerist ambitions.

At the capitalist, traditionalist end of consumerism, problems develop from the uneven distribution of consumer goods and services. This is shockingly evident throughout Latin America. Leaving a capital city is like crossing the River Styx. One leaves the technological-electronic age and returns to a world of nineteenth-century mechanics. Uneven consumer distribution creates enormous political strains and tensions. In short, modernization has to do with political leaders' strategies for the survival of social systems and not simply with historical tendencies.

Changes in basic commodities have profoundly affected all industrial economies. Different inputs and different outcomes come about from new empirical infusions. For one hundred years, economists lived in a neo-mercantilist world of their own invention in which finished products when bought fetched higher prices than the agrarian source materials did. It cost more for a television set than one earned by selling livestock. Now we have a world abounding in televisions, stereos, radios, and refrigerators, but an unsatisfied world demand for livestock and agrarian products.

We are confronted with a world of tradition *and* modernity: the rabbit's foot in the pocket of the industrial worker; and the submachine gun on the shoulder of the rural peasant. More important, the traditionalism that a society gives up may have been perfectly suited to produce quantities of foodstuff and livestock. Modernization may require new institutional arrangements which, in turn, compel a dangerous instability between politics and economics; and moreover between elites and the masses. Reality conspires against theory: this is why social science so often lags behind the real world.

The crucial problem of developing societies is not so much their innate inability to modernize but more the lack of institutional settings, regulative mechanisms, or normative injunctions upholding the new social system and capable of dealing with various policy differences as they emerge.[16] That is why aspects of traditionalism remain so firmly rooted during modernization. Conversely, a society that fails to incorporate traditional modalities into the modernizing process (however these terms are employed) is foredoomed to constant tension and opposition. The French Revolution of 1789 was supposed to begin all things anew, including the calendar, and to destroy particularistic and primordial symbols of local, ethnic, caste, or class groups. It soon found itself in the grips of totalitarian democracy. The manifest danger in the strict demarcation of past and present was that continuities and discontinuities in all forms of social relationships became evident. Legitimacy was surrendered, and the Revolution was overthrown. Thus, either too little institutionalization or too much innovation may lead to the breakdown of meaningful balanced development. Social change is volitional rather than deterministic.

The social sciences have had three developmental paradigms with three distinctive strategies. The 1950s was the decade of modernization as proclaimed by Walt W. Rostow, in which the United States was seen as the epitome of success defining all other social systems.[17] This idea has recently been extended to include a belief that competition and inequality are major sources of innovation.[18] The identification of modernization with Americanism broke down in the 1960s. In its place emerged faith in industrial development as the master ideology of science. This coincided with the first development decade, in which the United Nations became the voice for the Third World, and, in consequence, Third World science became an important factor in pushing research and development. The case for developmentalism was argued in its own terms, moving away from posing the issue as a choice between traditionalism and modernization to a more anti-colonial style. The 1970s witnessed the emergence of a demand for egalitarianism, at every level, from the most intimate relations between men and women to the most general relations between nations. Egalitarianism as a demand is the common ideology of have-nots. It has reorganized all our values, our sensibilities on what constitutes the just and the good.

The notion of egalitarianism as a societal goal has become linked to a growing awareness of the uneven distribution of world resources and this in turn promises to create a split within the Third World in the 1980s. One part of the Third World is becoming a modernizing sector, and another part a Fourth World, a dependent and backward sector. This Fourth World is increasingly exploited by advanced sectors of the Third World no less than by the First and Second Worlds. Concepts of traditionalism, modernization, and industrialization are meaningful only in the context of the empirical realities that all developed nations share: energy shortages, environmental problems, and egalitarianism as an international demand. The character of social science paradigms must keep better pace with the changing realities of the developing world. Meanwhile, let us summarize the present state of the art of developmentalism.

As a technical concept, development rests on a predisposition to consider advancements in science and technology as indicators of the way modernizing individuals overthrow the dross of tradition. But, sadly, moral changes operate on a very different, and slower, timetable. The technological miracles of the twentieth century have repeatedly turned into the nightmares of the century. To recall Hiroshima, Auschwitz, or the Gulag archipelago is to appreciate the degree to which the modernization of technology does not necessarily emancipate mankind from forced labor or arbitrary death. It would be a mistake to deny the validity of modernization as a conceptual device for the study of the developmental process. But it is a far greater mistake to see modernization as identical with development. Modernization may highlight certain essential contradictions within the developmental process, but it cannot seriously be viewed as either the essence of development or the motor force behind democracy. Modernization extends human capacities and horizons in two directions: to perform social good and to perpetrate unmitigated evil.

3

Social Planning and Economic Systems

The Political Context of Social Planning

The antecedents of social planning for whole systems have strangely contradictory histories within socialist and capitalist nations. Planning became the essential touchstone of the post-revolutionary Soviet regime. The ability to predict the future and organize the present to reach specific goals was viewed as a fundamental law of socialist society that distinguished it from capitalist society. Some critics refer to the Soviet Union as the epitome of an over-managed society:[1]

> The productive forces of socialism differ basically from the productive forces of capitalism in their social form, although they have much in common with the latter from the material and technical standpoint. The new social system has opened up for them great additional possibilities and advantages; a planned and rational use of productive capacities, greater efficiency in the organization of production, balanced development, higher rates of economic growth, unlimited overall technical progress, new incentives for labour, and the creation of a new man.[2]

For a considerable time, planning on a national scale, certainly as it relates to industrialization, was viewed within Anglo-American culture as a veritable harbinger of the communist menace, an example of what happens to a society lacking in freedom. A powerful anti-planning and anti-utopian literature sprang up and quickly took hold, even before World War II came to an end. In Friedrich von Hayek's words,[3] planning represented the "road to serfdom," while for Karl Popper[4] planning incorporated the worst features of social engineering and led to a "closed society." Even a hint, as in the work of Barbara Wootton,[5] that there could be "freedom under planning" was met by a barrage of denunciation. Milton Friedman[6] put the matter in blunt economic words: "Collectivist economic planning has interfered with individual freedom." More recently, Robert Nisbet made the same point in sophisticated sociological terms: "Large-scale government, with its passion for egalitarian uniformity, has prepared our minds for uses of power, for invasion of individual privacy, and for the whole bureaucratization of spirit that Max Weber so prophetically identified as the disease of modernity."[7]

While laissez-faire European and American conservatives were denouncing the planning ideal, American planners, often quite conservative in their economic and political outlook, went about their business oblivious to and unconcerned with the ongoing ideological barrage. Within an Anglo-American context, planning was institutionalized, not via Marx, but rather via John Maynard Keynes. Planning in America became at least as extensive, if not as intensive, as planning in the Soviet Union. If the Soviets were primarily concerned with planning in the industrial sector, Americans employed planning in urbanization, education, transportation, neighborhoods, and almost all areas outside the world of industrialism.[8] "Free enterprise" and the "market economy" were allowed to flourish unhampered exactly in those areas where Soviet planning was most intensive, in both the factory and industrial work settings. In more recent decades, beginning in the 1960s, business has gravitated to a limited planning model. But such planning is integrated neither at a national nor even industry-wide level; rather planning is seen as exclusively within the domain of the individual firm. Indeed, long-standing anti-monopoly legislation also serves to inhibit social planning on a larger scale. Thus, a peculiarly distorted vision of planning prevailed in both worlds, each protecting a charade of freedom—be it "democratic centralism" or "free enterprise"—which assumed that industrialization and modernization necessitate the guidance and forecasting of needs, performances, and growth potentials in a world of scarce resources.

The specific mechanism that brought planning to the fore in the United States was the New Deal, followed by World War II and, in rapid succession, the Marshall Plan and the War on Poverty. Each of these efforts used planning almost as an afterthought, to deliver on political promises. The goals were announced, the policies enumerated, and the plans followed in helter-skelter fashion. Planning in the United States was never an ideological imperative as it was in the Soviet Union, but instead a functional imperative, situationally determined.

In his "vignette" of the New Deal, Lepawsky indicates how many policy formulators, program administrators, and planners coalesced in the Roosevelt administration, never again to be parted:

> Despite devastating criticism from the opposition, planning burgeoned and virtually became a national policy in its own right during the New Deal. Often in disarray, but always dynamic, this system of policy planning and program planning embodied the aims and brains of the Roosevelt administration. It was this powerhouse of intellect and interest, experience and experiment, which, by 1939 when the New Deal stood at its midpassage, constituted the core of the planning apparatus. Reciprocally the planning apparatus had by 1939 virtually become the cortex of the Brain Trust.[9]

So firmly embedded had planning become by the end of the 1930s that the movement from warfare to welfare was accomplished with hardly a bureaucratic ripple. The continuum between the U.S. Departments of Defense and of Health, Education, and Welfare continued into the postwar era—cemented by the planning factor.

The profession of planning originated in connection with accounting, engineering, and finance, and operated within a bureaucratic context. Planning came into its own during the 1960s, when planners had as their essential task to deliver on the long-range rational goals of society. Planning in the United States, far from being an ideological imperative demonstrating the superiority of the free enterprise system, was in fact an essential handmaiden of the industrial and urban complex. If ever a demonstration were required that practice has its own imperatives, planning would surely be a perfect example. Planning resources depend on the inexorable complexities of urban industrial life far more than upon the rhetoric of liberalism or radicalism. Whatever else planning is, few can doubt its essential place as the centerpiece of twentieth-century bureaucratic administration.

Postwar planners came to recognize the risks involved in a highly aggregated definition of the public interest. Planning was often thought to be in opposition to community welfare. The rationality of the planning syndrome was said to involve irrationalities at the level of implementation.

> Our concepts of optimality, our focus on an abstract welfare function, and our concern for an illusory greater good (or "public interest") is brought into serious question. Planning is being challenged more and more, not on its service to an overall public, but rather on the differential and distributional aspects of its results affecting particular publics.[10]

By the 1970s, it had become part of the new conservatism to suggest that communities, rather than regions or nations, should be viewed as the basic systems unit, directing greater attention to the needs of the local community.

> One way to do this might be to recognize *communities* as the fundamental systems units, and attempt to build up regional service systems from community-based modules supplying urban services. By thus inverting the planning strategy, it may become possible to adequately represent not only processed knowledge, but, along with it, the personal knowledge of community and neighborhood impacts in urban system planning.[11]

The difficulty is that even with a more expansive vision of constituencies, the planner seems unable to get beyond an interest-group model into a general theory that would in fact entitle planning to be considered a social science.

What makes "liberation planning" or "participating planning"[12] difficult, if not impossible, to achieve is the constituency that planning agencies serve. Market values have their own laws, and planners serve economics. This is just as true in a Third World context as in more developed societies.

> As long as discount value dominates economic calculations in resource allocation decisions, it appears difficult indeed to opt for wildlife protection or aesthetic urban design—unless, and

> only unless, those can be rendered in favorable quantitative terms within a context of discounted value. We must show, in that case, sufficient near-future, if not immediate, quantifiable pay-offs for keeping the fish and birds at the hypothetical pond. The birds and fish must be valuable for killing insects which bother people or for providing recreational sport, or some such utilitarian benefit. This view decrees that if the wildlife cannot serve man some way, it has to go. And wildlife has been going, as ponds and meadows are filled, streams are polluted, and ecosystems are disrupted. Seldom, though, can discount calculations justify otherwise.[13]

The degree of restraint one encounters in present-day planning literature contrasts markedly with the euphoria characteristic of planning literature during the sixties. The themes of "brave new world," "progress," and "change" were so overwhelming that one text insisted that we were "beyond debating the inevitability of change." The notion of planned change was considered so central that "no one will deny its importance."[14] Amidst movement and rebellion and new demands for racial, ethnic, and sexual equity, one can well appreciate such a sense of confidence. Under such circumstances, there seemed no question that social planners, policy makers, and social scientists would come together in a grand assault against the establishment. In the move toward planning, neoconservative attacks were dismissed for having so wide a set of assumptions about tradition that they betrayed the very variety of the authoritative traditions to which they appealed. Why a similar argument could not be adduced against pluralistic liberalism and its multiplicity of theories of change was not made clear. Yet, it is clear from the perspective of only a decade ago that the much heralded unification of social planning and social science has not taken place. Their shared participation in the glories of policy making have likewise not materialized. Why these anticipated outcomes have not occurred is, in large part, the subject of this chapter.

Social Planning and Social Science

Social planning and social science, far from being "handmaidens" in a common concert, are in fact distinct in themselves and for the most part profoundly antithetical in their missions. The common sense of planning rests on the maximum utilization of resources and perfect equilibrium of economies. The common sense of social science rests on maximum equity and liberty, which in turn entail producing an excess of supplies to satisfy demands. The bugaboos of planning—waste, unused resources, and inefficient organization—may be seen by social scientists as preconditions for a decent society serving the public interest.

There is a considerable body of literature informing us that social planning is in transition, crisis, disarray, transformation, etc., but precious few planners have paused to give us a workable definition of planning. About the closest thing to a definition to be found in a basic text on the subject of urban planning is that planning represents an "action-producing activity which combines investigation, thought, design, communication, and other components. But in another sense, planning is a special kind of pre-action action."[15] It is by no means unkind to note that this valiant effort at definition is viewed even by its progenitor as "two steps removed from an actual environment-shaping activity," and is at best "an efficient substitute for trial and error regarding the relations of wholes and parts." This strongly subjective definition of planning in terms of a process of feeling, knowing, and acting invites grave doubts as to its structural status, or placement within a social science cosmos. Yet, even the most rigorous social planners characteristically define the field in highly operational terms, clearly dependent upon constituencies rather than concepts, doers rather than demographers, and systems rather than organizations.

Despite the noteworthy efforts of some to create a "linking process"[16] between physical planning and social planning, it is evident that as definitions of a "decent home and suitable environment" came to signify moving to suburbs and high income areas rather than systematic urban redevelopment, the linkages became more apparent than real, and old ideological hostilities reappeared. Demands were made that the "inequalities and pathologies of the urban low-income population must be eliminated before the attractive, efficient, and slumless city for which physical planners are striving is realized."[17] Further, just as "slums" and "ghettos" were subject to a labeling process that magically transformed them into "communities" and "neighborhoods," so too did the distinction between physical planning and social planning become highly subjective. Whether uniform building codes based on performance standards are actually a function of physical planning or of social planning very much depends on who is establishing the codes and standards, and negotiating them to what purpose. Again, the absence of simple guidelines for the concept of planning in general is remarkable.

The essentials of planning appear simple enough. Planning is charting future developments in terms of present tendencies and trends, which, adjusting for changes in technology and the economy, can predict lifestyles, industrial growth, and real wealth of a future population. To extrapolate further from the literature: planning can be either short range or long range. It can be concerned with planning a meal for tomorrow, or planning soft landings on Mars ten years hence. Planning, in short, at some level is forecasting; presumably, the better the forecast the more accurate the plan, and in this sense the more like engineering than science.[18] One might also add that it is more like art than engineering.

Deriving indicators or benchmarks for proving or disproving the worth of a plan becomes the essential task of social accounting, which in turn relies upon social indicators for information. Planning involves maximizing rationality. It presumes that targeting goals is eminently feasible, even necessary, and probably desirable, whatever idiosyncrasies may appear in human behavior in the first instance, economic systems in the second, and technological innovations in the third. This may be a not completely orthodox view of planning, and it certainly is open to challenge, but it seems fair to say that for most people in the business of planning, most, if not all, of the foregoing is considered a precondition for success.

Even those statements by planners that recognize the need for precision, for "forced choices" employing the game analogies of the behavioral sciences, easily identify social planning with the public interest. Planners are urged to strengthen the ability of political leadership to respond,[19] and the social planning field finds itself under assault and in a weakened position. To raise the cry of public interest or betterment of professional standards in a shrinking market is not to address the problems faced by social planners. Advocacy of what, and planning for whom, remain unanswered questions buried beneath the rhetoric of the public interest.

In a sense, the central political role of social planners made them unlikely candidates as social science performers. As Rabinovitz[20] indicated, apart from forces beyond their control, social planners do assume various political guises: the roles of technician, broker, and mobilizer being paramount. In addition, the social planner must sometimes initiate and at other times veto political directives. But these multiple roles, brought to fruition in an authoritarian context, represent far less (or more) than a scientific appraisal of circumstances or situations.

Planning remains a process far different from what planners describe. In point of fact, only rarely do planners define the planning situation. It is instead defined by the political system; hence the goals set are very often outside those indicated by the planners themselves. Without getting into the literature on the struggle between bureaucrats and politicians, it should be evident that planning involves a huge mobilization process. To set goals is the task of the political system; to implement goals is the task of the planning community. However, the political system sets its plans in terms of parameters having little to do either with the available technology or the available bureaucracy. Political life depends upon events external to industrial planners. When Stalin declared, in 1931, that Soviet industrialism would have to reach the level of the West in ten years or face utter defeat in another world war, he was accurately assessing the international situation. It then became the task of planners to set specific Soviet industrial goals. It also became the task of the police and the secret police to make sure that those goals were achieved according to plan, whatever the human costs.

Those who celebrate the success of Soviet industrialism almost, but not quite, mask the human costs involved in its realization. In Third World countries where police functions are minimal, equivalent planning results cannot be guaranteed. Thus, despite Fidel Castro's announcement of a ten million-ton sugar harvest, the net yield was only six million tons. The police force never operated with maximum force, and thus the political leadership was embarrassed. One might conceivably argue that by 1939 the New Deal planning system, in much the same way, had come to a dead halt. This might easily have led to the demise of the New Deal, but for the divine intervention of World War II, which provided new tasks for planners and new aims for the political leadership. This is not intended as a cynical mini-review of history, but as an indication that planning is not simply a technical chore performed by sophisticated technical personnel; it is in fact a janitorial mission in the service of the political system, with mobilization being the crucial element in the process.

There is implicit in the planning function a political morality. Whether it is stated in terms of a new society, a new man, or a new deal, the moral foundation of planning is absolute rationality. Without such rationality, forecasting clearly becomes an odd game. Hence the planner makes a strong investment in rationalizing human behavior, not because of any propensity toward rational concepts on the population's part, but because irrational behavior becomes a mode of destroying the realization of the plan. As a result, within a Western context, things like strikes, work stoppages, slowdowns, and sick days are thought of as vicious attacks on the plan—downright irrational forms of behavior—as if the purpose of the strike is to thwart the planner rather than to gain better wages, better working conditions, or better working hours. Of course, in the Soviet context, this attitude was stated more bluntly, in terms of those who would strike or slow down—in short, those who refused to receive the benefits of the plan by refusing to implement the costs of the plan.

National planning and the global planning ethic, far from providing a basis for radical behavior, were in fact sources of a new conservatism: an industrial conservatism, whether of a capitalist or socialist variety, which saw all shortfalls with respect to achieving goals as illustrations of the miserable nature of human behavior, of the willful, spiteful, and hateful character of people, of an irrationality that had to be destroyed. Thus, what started out as a new economic mobilization leading to a new society and a new man became in fact the captive of the oldest political network of all: dictatorship and rule from the top down. Never has the road to hell been paved with more honorable intentions.

As an effort to cope with this monster in the machine, the planners began to realize both by themselves and through the criticism they received from the social science community that planning is not all of a piece; or, better, should it be of one piece, it is terribly dangerous. One could conceive of planning in various ways.

Under contemporary capitalism there is private sector planning, whereas under socialism there is public sector planning. The problem here is that public planning has many options available to it that private planning does not. The public sector can lower taxes and provide rebate incentives, intensify inflation, or create new jobs. About the most the private sector can do is move the profit margin up and down as a means of inducement and enticement to buy or to sell. Hence it becomes evident that planning in a public sector is of necessity quite different from planning in the private sector and has more possibilities for realization as well as more possibilities for abuse. As Weidenbaum points out:

> There are fundamental differences between business and Government planning. Essentially, we are dealing with the difference between forecasting and reacting to the future, and trying to regulate it. Corporate planning of necessity is based on the principle of trade—attempting to persuade the rest of society that they ought to purchase the goods and services produced by a given firm; the controls that may accompany the plan are internally oriented. In striking contrast, the Government is sovereign and its planning ultimately involves coercion, the use of its power to achieve the results that it desires. Its controls are thus externally oriented, extending their sway over the entire society.[21]

Planning takes place at local, regional, national, and even international levels. Clearly, the question of planning at the international levels is more problematical than planning at the local level and is fraught with many more possibilities of failure since many uncontrollable variables are involved, not the least of which are the various interpretations to which international law is subject and the unpredictable nature of international conflict. Thus, as one moves from local planning, which offers maximum possibilities for forecasting, to national and international planning, one is moving from a more successful to a less successful sort of planning.

There is also the question of the advantages and disadvantages of long-range and short-range planning. It might be the case that short-range planning will have greater propensities to precision and exactitude than long-range planning, since it involves fewer budgetary decisions. Theoretically, a five-day plan should be easier to realize than a five-month plan, which in turn should be easier to realize than a five-year plan. On the other hand, time may work the other way, since failures in one year may be made up in a following year, whereas with a short-range plan the possibility of make-up or correction is almost entirely lacking.

A still more advanced aspect is planning from the top down versus from the bottom up. By "from the bottom up," I have in mind a kind of planned self-management characteristic of Yugoslavia and parts of Italian industry, in which the actual planning is done by factory workers. Likewise, bottom-up planning can be similar to recommendations made by condominium developers who determine for themselves what the wages and salaries of management ought to be, what improvements can be made, and under what forms of supervision. Top-down management is so well known it requires little illustration. But here, too, there may be a problem of apiority. It may well be that top-down management is superior because of the waste involved in bottom-up management and the conflict engendered by everyone having his say.

There are residual and large-scale questions concerning goals—increase in equality, for example, versus increase in profitability. These may involve issues ranging from environmental controls to housing start-ups. Compliance with all of the rules and regulations of the Environmental Protection Agency might substantially reduce equity, which is the ostensible goal of such an agency. Equity may be frustrated by the absence of economic growth per se.

These types of planning problems—conflicts between private sector and public sector, regional versus national, top-down versus bottom-up, long-range versus short-range forecasting, social welfare goal orientation versus profitability orientation—are in fact empirical problems that can be analyzed only within a context of social research.

The planning field provides data for a rather interesting fourfold table. Some political radicals see planning as the very touchstone of democracy, while others argue that social planning is little more than a statist-inspired, self-serving instrument for maintaining class and race differentials. Among political conservatives, some see planning as blurring the hard and firm necessity to respond to the economic marketplace, while others see planning as a final barrier to anarchism in our times and a necessary component of law and order. Hence, those expecting neat correlations between planning and ideology are likely to experience serious frustration. For this reason, a functional rather than an ideological account of the relationship between social planning and social science is required.

Given the fact that relatively few planners are either trained in or make their living through the social sciences, it is hard to accept the well-intentioned belief that of all the client-serving professions, planning is among those which call on the greatest depth and diversity of social science disciplines; and it is equally difficult to view the planner as a social scientist turned practitioner in the arena of public policy. Beyond the demographic distinctions is the central fact that any social science must be more than a profession; it further requires a critique of its own practice, as well as that of public policy.

The rhetoric and professional identity of planning have increasingly moved in the direction of social science. A study of M.I.T. graduates, for example, indicates that in the early 1960s the background of planning personnel was largely engineering, whereas by the end of the decade, and continuing into the 1970s, architecture and the social sciences were most often mentioned as the undergraduate majors of prospective planners.[22] This report indicates that fewer than 20 percent of planners have the university as their primary job location, or combine the categories of academic researcher and teacher as their primary role identity. Another recent study of state planning directors reveals even more emphatically the gulf between the profession of planning and the various sciences of society.

> Certain attributes of state planning directors—their background and their office—are correlated with the relevant performance of state planning activities. Their level of education appears to be irrelevant; the type of education is only slightly related; but their career background is highly significant, especially in terms of prior work in management, political experience, and mobility. And the closer the relationship of the director and the governor, the more likely is the agency to be performing activities. These

> conclusions do not enhance the prospect of seeing state planning become a science which can be learned in school. They suggest an art which must be learned by experience.[23]

What is suggested by this and similar data is not simply the craft-like nature of planning but its political context, which makes any impulse for autonomy and self-criticism a special responsibility of social scientists, not so much by empirical design as by *fiat*.

The increasing number of reports of social science advances and planning applications emphasizes the gap that remains between social science and social planning.[24] But these reports concentrate on "positive" criteria, i.e. methodological, theoretical, and institutional breakthroughs that are incorporated into planning practice. It is a measure of planning myopia that breakthroughs from operations research to cost-benefit analysis are considered only in terms of growth by stages. The dynamic of social science discovery, the interpenetration of research and criticism, and the consequent emergence of new paradigms tend to be ignored, or even suppressed, in the mechanistic rendering of social science innovation by the planning community.

The various summaries of the impact of the social sciences on planning reveal an underlying conflict between those who are in search of formal generalized models of planning and those who view this effort as mechanistic and pernicious. Only recently has the search for social science antecedents turned away from the customary sources to the contributions made by phenomenology, the new philosophy of language, and symbolic interactionism within sociology: a set of antecedents very different from what one had been accustomed to find in the planning literature. But this in itself would indicate both the dubiousness of a general theory of planning and, of equal importance, the degree to which the literature of planning remains firmly linked to theoretical and methodological developments within the social sciences.[25]

The recent tendency to make planning more responsive to a "public philosophy" illustrates the critical difference between a planning standpoint and a scientific standpoint. John Friedmann quite persuasively argues that the end result of the reform movement within planning was a statement of public interests that was little else but a statement of the interests of the planners themselves. He seeks to remedy this situation by relabeling public interest as public good, and by changing the equity emphasis from opportunity networks to results.[26] But both the conservative sociologist Robert Nisbet and the liberal sociologist Herbert J. Gans indicate the concern for a breakdown in community dominion. Nisbet indicates that "there is no likelihood of our achieving cultural or social pluralism, of genuine and creative localism, of regionalism, so long as present tendencies toward centralization of power, administration, and function go on and on."[27] And Gans, while sharing Friedmann's political goals, nonetheless concludes by stating that "even if all planners agreed to be in the public interest, it would still have to be achieved through political struggle."[28]

An equally devastating critique of the idea that planning somehow represents a unitary community interest, or an objective social science, is contained in Frances Fox Piven's blunt statement on planning and social class:

> Planners were committed to the values of growth and development, and to the economic and political interests in the city that prospered through growth and development. Planners did indeed take sides, and they took sides with the powerful, with the city builders. It was also not true that planners played a large role in the decisions that shaped the form of our cities. To say that city planners served the city builders is not to say that they made the city building decisions, or even that the plans they prepared were of significant influence. The key decisions, the decisions that accounted first for the huge concentrated agglomerations of our older industrial cities, and then for the subsequent evisceration of these cities as capital moved to the outlying rings and to new cities of the south and west based on oil, autos, electronics, and aerospace, were never decisions embodied in any plans made by planners. Compared with the formative influence of capital investment decisions, planners and their plans were mere shadow play. At most, planners only struggled to service the cities built by private capital with the support of public capital.[29]

The point is not to castigate social planners; indeed, even severe critics hold out the prospect of a different kind of planning based on the new urban constituencies of poor whites, ethnics, and blacks. Nonetheless, it would be folly of the misanthropic type to assert an isomorphism between social planners and social scientists, given conditions that clearly point in an opposite direction.

Measuring Success and Failure in Planning

There is a national tendency to think of social planning and social science in strictly cumulative and interactive terms. John Friedmann characterizes basic discoveries that are implemented by planners, and Lawrence Mann discusses the same subject in terms of the continuous shortening of lag-time in the implementation of social science findings (theoretical, methodological, and institutional) in the planning process. But this consensual view assumes the purely cumulative, aggregative nature of social science and ignores its critical functions. Beyond that, there are grave doubts that the basic breakthroughs are really as stated. It is problematic whether theoretical approaches represent an institutional breakthrough or a waste of public resources. Those who have studied this matter have pointed out that "think tanks" work in areas where a clear consensus exists (i.e. racial equality in housing and education) and fail miserably when addressed to areas in which a wide dissensus exists (civil action to counter guerrilla movements, or, as in the *Pentagon Papers*, studies addressed to the conduct of an unpopular war). Admittedly, it is flattering for social scientists to read that social planners are absorbing their literature more rapidly; it legitimates the distinction between an applied craft and a scientific profession. But it fails to address the degree to which issues in planning are absorbed, if at all, by social scientists; and, more pointedly, whether the absorption process, to the extent it operates, may serve to weaken the critical resolve of social researchers.

A considerable amount of social planning, especially urban planning, involves developing appropriate models and establishing suitable criteria. Hence the research burden shifts from goals to methods. The burden is on information systems, data processing, how planners can locate manufacturing supplies or proprietary software or federal agency-sponsored software, accept and deliver data in a variety of forms, improve procedures, develop administrative records as new data sources, and collate data collection of dates and time intervals.[30] While emphasis on means rather than ends is perfectly acceptable for an applied discipline, the critical test remains the implementation of modular approaches within the intrusive complexities of the ordinary world rather than how they fare during the initially delicate and protected trial period.

Planners' commitment to formal experimental protocols and models unfortunately seems to fade precisely when this implementation phase is reached. As a result, there is little effort made to determine whether a plan worked or did not work, and if it did not, whether this failure was due to the choice of goals, the mix of strategies, or specifically, the configuration of tactics.[31] Examining the planning literature, one is struck by how little effort is expended on questions of what went wrong, and how much instead on model construction, despite the general recognition that much does go wrong when plans are implemented in real-world terms.

Social science must point out that planning involves planners. To carry this beyond the level of tautology, social science must include an analysis of social planners: problems of occupations, income, and stratification. How people within this occupational group relate to one another becomes a central factor in the success or failure of a planning mechanism. The social scientist can provide the self-consciousness, or perhaps the consciousness of self, the planner so often lacks. For example, one sociologist has pointed out how the gulf separating master planners and advocacy planners came about: "The master planners were advocates for themselves, for the city's business interests, and for the upper- and upper-middle-class residents of the community. Indeed, modern advocacy planning, which seeks to represent the interest of the poor and the black community in the planning process, has developed precisely because comprehensive planning has largely ignored their problems, goals, or needs."[32] Without prejudicing the merits of such an approach, it clearly places the planner in a context of stratification and demographic information. For example, what are the background variables of the planners; from whence do they derive? Unless one believes that background variables bear no relation to behavior, this is exceedingly vital information, first, in distinguishing the planning community from the social science community, and, next, in separating out the planning community from the political establishment, itself a vital undertaking which social research can perform for social planning.

Because of the essentially dependent role of social planners and their lack of scientific autonomy, they perform the same sorts of service roles in socialist countries as they do in capitalist ones. In the rush for industrialization, the kind of polluted environment socialist planners thought was a result of aggressive capitalist development occurred in socialist nations as well. Thus, it is not merely service to the market that becomes a problem but service to the state. Which economic system is the harsher taskmaster can be left to the imagination of those familiar with comparative economic systems.[33] Under socialist planning, the very survival of an independent social science is threatened in the rush to convert the chaotic present into an orderly future. Criticism becomes suspect, and constructivism becomes the marching order of the day.

Aaron Wildavsky has summarized the role of planners with unusual clarity. Once planning is viewed within a perspective of society, and society itself is considered a system of power, then the actual relationships of planning, budgeting, and reform become quite clear:

> Where disagreement over social goals or policies exists, as it must, there can be no planning without the ability to make other people act differently than they otherwise might. There would be no need to plan if people were going to do spontaneously what the plan insisted they do authoritatively. Planning assumes power. Planning requires the power to maintain the preeminence of the future in the present. The nation's rulers must be able to commit existing resources to accomplishing future objectives. If new rulers make drastic changes in objectives, the original plan is finished. . . . Planners are spenders. Their raison d'être is economic growth. Typically they underestimate spending and overestimate revenues to leave room for investments they believe are necessary for accomplishing the goals they wish to achieve. Planners are natural allies for large spending departments whose projects planners believe desirable for natural enemies of financial controllers who want to limit expenditure. If planning could not control budgeting, budgeting might yet become a form of planning. By expanding budgeting to include planning, the same goal could be achieved from a different direction. The idea is marvelous; if planners suffer a power deficit, they balance their books by becoming budgeters, who have a power surplus. Budgeters are powerful but ignorant;

planners are knowledgeable but powerless; what could be more desirable, thought the proponents of PPBS, than combining the virtues of both classes by making budgeters into planners.[34]

The trouble is that social planners no longer like the idea of seeing themselves as bookkeepers or accountants. This is where the problem of planning as a social science is joined. It is evident in the work of people like Wildavsky that the analysis of social planning is itself open to social research, whereas the very nature of the planning process limits that self-awareness or even the possibility of self-criticism. Social planning is connected to power, and science to knowledge. To see the requisites of planning simply in terms of knowledge is to doom planning itself.

The most precise way to underscore the difference between social planning and social science is by example. The planner must assume optimum rationality. This is interpreted as optimum usage of a facility. Economic theory, however, if not always directed toward the public interest per se, might assume that the functions involved in this kind of planning optimality—for example, the constant queuing to get service, whether it be a retail facility, the cinema, or a bathhouse—may make the very notion of planning dysfunctional from the point of view of encouraging incentives and hence economic growth. This is exactly what the Soviet economist, Yevsei Lieberman, pointed out in relation to Soviet planning. In a delightful essay entitled "Waiting for a Bath—and Just Waiting," the essential difference between social planning and science is made painfully evident:

> Why must we, for example, stand in queues for hours to buy railway tickets, particularly during the vacation season? Frequently we do this not because the capacity of our railways is insufficient but because of the hidden workings of the theory of so-called "full-capacity." It is considered profitable to have a railway timetable in which not one single seat remains empty. And the bath—excuse my frankness—what's it like to visit a public bath on Friday or Saturday? First you must stand in a queue, because the "washings per person" are planned on the basis of "full bath benches and tubs." From the standpoint of public interest, such unoccupied "optimum" facilities are also suitable to movie houses, on trains, in post offices—briefly, in all public-service establishments. There is no other possible solu-

tion. If there is to be no queuing, then supply must everywhere exceed demand. Only in this way can we provide high quality products and conscientious service.[35]

Following through on the Soviet situation, there has been a crescendo of criticism concerning state planning committees and the Supreme Council of the National Economy. During that remarkable period of ideological thaw in the early sixties, a host of criticisms was made of planners. Their essential thrust was that, although they provided for the opening and completion of the plan, planners failed to address themselves to the placing of orders for equipment for construction projects; construction costs had to be revised upward for almost half the projects, despite the fact that many of the designs had been rated of high quality. Wildavsky's charge about American planning is doubly true for the Soviet system. There is a strong tendency to equate maximizing results with minimizing costs. Deputy G. I. Popov makes this point strikingly evident:

> In recent years the construction workers of Leningrad and other provinces of the country have repeatedly proposed the introduction of a method of financial accounts for fully completed projects only [not to make payment until a project is complete]. Such a practice stimulates better work on the part of the construction organizations and contributes to the quicker opening of new structures. The government has approved this initiative, but through the fault of USSR State Planning Committee this completely progressive method of financing construction work has not yet been introduced.[36]

This was followed by another report summarizing the shortcomings of Soviet planning.

> In our opinion, the chief shortcoming is the unscientific, arbitrary planning that has taken root in many cases. Departmentalism and local allegiances lead to the fact that the construction of more and more new projects is launched while at the same time already built production premises and units are being poorly utilised and technological processes are not completely operational. This occurs because there is inadequate responsibility for the working out of the technological part of the designs for enterprises. Moreover, the compilation of the plans for

> capital construction is delayed until the end of the year. Furthermore, at the outset the plans are inflated, and then they begin to reduce and tighten them. Therefore decisions on inclusions in or exclusions from the draft plan are frequently made hastily. Where now is your meticulous economic verification of the designs and your genuinely scientific planning![37]

It is crucial to outline the constraints on socialist planning in terms of the breakdown of innovation and the resulting inroads on efficiency. In the Soviet model at least, the greater the degree of planning, the less the degree of innovative capacity; and the higher the degree of efficiency, the lower the degree of planning. In recognition of this, the Soviet bloc has introduced market mechanisms such as incentive systems, differential pay structures, the attenuation of discrimination against private sectors, and the reintroduction of notions of profitability and a market network. The limits of planning are nowhere made plainer than within the Soviet framework, the result of which is a borrowing of foreign technology and the intervention of foreign firms in order to overcome the breakdown of internal innovation. The strains between party absolutism and economic relativism are such as to make impulses toward efficiency and innovation subject to communist politics and ideology.[38] Under such stress, purchasing overseas research and development permits weaknesses in centralized planning to be papered over.

Precisely the structural deficits of the master plan and of centralized planning have led to a drastic breakdown of Soviet innovation and efficiency at crucial industrial points, especially in areas of older industries where the fulfillment of plans does not depend upon technological breakthroughs. Berliner summarizes the strains between planning and innovation neatly in terms of the transformation of capitalism and the "invisible hand" into socialism and the "invisible foot":

> Adam Smith taught us to think of competition as an "invisible hand" that guides production into the socially desirable channels. By a curious ideological confluence both Adam Smith and the designers of the Soviet economic structure had in mind the smooth allocation of resources under a basically unchanging technology. Central planning may be regarded simply as a visi-

> ble form of the same guiding hand that operates invisibly in capitalism. But if Adam Smith had taken as his point of departure not the coordinating mechanism but the innovative mechanism of capitalism, he may well have designated competition not as an invisible hand but as an invisible foot. For the effect of competition on innovation is not only to motivate profit-seeking entrepreneurs to seek yet more profit but to jolt conservative enterprises into the adoption of new technology and the search for improved processes and products. From the point of view of the static efficiency of resource allocation, the evil of monopoly is that it prevents resources from flowing into those lines of production in which their social value would be greatest. But from the point of view of innovation, the evil of monopoly is that it enables producers to enjoy high rates of profit without having to undertake the exacting and risky activities associated with technological change. A world of monopolies, socialist or capitalist, would be a world with very little technological change.[39]

In adopting central planning the Soviets achieved the benefits of the invisible hand but lost the advantages of innovation. Worse, they were not able to back away from mistakes—i.e. not properly adjusting for inflationary rates or breakdowns in shipments and supplies—since error factors were not built into the planning network itself. Thus, under capitalism, even large-scale errors, let us say in the automobile industry or in the chemical industry, as well as development and promotional costs can be absorbed by more efficient sectors far more readily than in the Soviet system, where the organizational structure protects producers against losses from both their own unsuccessful innovations and the successful innovation of others. The lag between planning and implementation may itself become a major obstacle to innovation. Again, the central point, in terms of our own task, is to show how clearly and distinctively the process of social planning differs from the process of social science. For only a science of economics could make these points about planning without fear of either contradiction or destruction.

Nor should this be viewed as simply a Soviet problem. In point of fact, if one turns to urban planning in the United States the same kind of optimal rationality produces a similar kind of public inequity, and at times results in anti-federal "pathologies." Herbert

Gans indicates that the upper-class origin of planning in America made such lower-class pathologies inevitable.

> City planning grew up as a movement of upper-middle-class eastern reformers who were upset by the arrival of the European immigrants and the squalor of their existence in urban slums and the threat which these immigrants, and urban-industrial society generally, represented to the social, cultural, and political dominance the reformers had enjoyed in small-town agrarian America. As reform groups and businessmen gave city planning increasing support, it became a profession. Its physical emphasis naturally attracted architects, landscape architects, and engineers; these developed planning tools that were based to a considerable extent on the beliefs which the movement had accepted. The inqualities and pathologies of the urban low-income population must therefore be eliminated before the attractive, efficient, and slumless city for which physical planners are striving is to be realized. When the latter can be persuaded to the validity of this concept, it may be possible to achieve a synthesis of the so-called social and physical planning approaches to create a city-planning profession which uses rational programming to bring about real improvements, not only in the lives of city residents but also in the condition of the cities themselves.[40]

The same fears about planning are now being voiced in an American context across the stratification axis—specifically in terms of weaknesses in federal regulation as a whole, rather than planning mechanisms in particular. The isomorphism between Soviet and American social science concerns can hardly be more clear than in the following description of the costs of federal regulation:

> Federal regulation adversely affects the prospects for economic growth and productivity by laying claim to a rising share of new capital formation. This is most evident in the environmental and safety areas. It is revealing to examine the flow of capital spending by American manufacturing companies just prior to the recent [1972] recession. In 1969, the total new investment in plant and equipment in the entire manufacturing sector of the American economy came to $25 billion. The annual totals rose in the following years, to be sure. But when the effect of inflation is eliminated, it can be seen that four years later, in 1973, total capital spending by U. S. manufacturing companies was no higher. In "real terms," it was approximately $26 billion

> both in 1969 and 1973. The direct cost of government regulation, a topic rarely studied, is substantial. The number and size of the agencies carrying out federal regulations are expanding rapidly. The administrative cost of this veritable army of enforcers is large and growing. The costs of government regulation are rising far more rapidly than the sales of the companies being regulated. Regulation literally is becoming one of the major growth industries in the country. But this represents only the tip of the iceberg. It is the costs imposed on the private sector that are really huge, the added expenses of business firms which must comply with government directives, and which inevitably pass on these costs to their customers.[41]

The central contribution of social science to the world of planning has to do with levels of planning. It may well be that the social-indicators approach, or the ecological movement generally, contains within itself a certain rebellion against the content of planning. It may also be that a revolution of falling expectations that comes about through a worldwide redistribution of the sources of wealth will serve as a limiting device to national planning. The social scientist also lives in a world of interests rather than a world of production. These interests may have a sharply limiting effect on the viability of planning, even if the plan is entirely rational and empirically feasible on fiscal grounds. There might be a question of interests that have been overlooked—racial, class, ethnic, sexual, or just plain interests of preservation instead of development.

The evaluation of the plan cannot be left to the planners any more than it can be left to the politicians. Left to the politicians, all planners would be shot for failure. Left to the planners, all politicians would be eliminated for demanding unreasonable economic goals from their modest skills. In such a context, the social scientist is uniquely capable, if not destined, to discuss and decide upon the efficacy of specific plans. We need merely turn to several instances from the world of Soviet planning and its failures, and to the world of American planning and its failures, to indicate how essential is the role of social science as honest broker in this tension-filled context of planning and politics.

When examining the literature on planning, whether it be local regional, or national, whether it be Marxian or Keynesian, or American or Russian, one notes a strange desire to solve the problem of planning within a planning context. Yet, those who contrast populism with planning often hold out few alternatives. It is as if these writers think that elliptical slogans such as "power to the people," one abstraction, will overcome the "power of the planners," another abstraction. Anarchic resolutions might be premature if not downright destructive. Anarchism and conservatism start from different philosophic premises, and each carries with it penalties for massive sectors of the American population. In such a context, it behooves the social scientist not to become captive to anti-statist ideologies any more than to state authority.

An impressive study on the relationship of justice and planning indicates by means of a "claims matrix" that the choice for the planner is rationally uni-dimensional, that in fact, in a hypothetical situation of clean water requirements, the needs and wants of the environmental party might be quite different from the needs and wants of the trade union party, if we can call it that. To have water clean enough for swimming may signify a level of unemployment that is unacceptable, whereas to have water clean enough for boating, but not for fishing or swimming, may involve zero unemployment yet not satisfy the advocates of clean water. Berry and Steiker properly point out that

> When the claims of two or more groups conflict, efficiency or net benefits are less relevant criteria. In such a situation, maximizing net profits will not prevent, and in fact may require, having certain groups subsidizing others and perhaps suffering severe and damaging losses. The key consideration in these issues then becomes the distribution of costs among the different groups.[42]

One notes even in this sophisticated study that decision making under such circumstances is really not a matter for the planner. They refer to such decision making as a Sisyphean task, and again appeal to that mythical "general good," in this case called the "publicly argued and the publicly decided," which in point of fact indicates the pragmatic limits of a planning effort. In other words, the fairness doctrine is really outside the purview of the planning mechanisms. Interestingly, in this environment/economy tradeoff the situation is so structured that any new employment opportunities created by firm ecological standards is not entertained; nor is there discussion of the possibility of solving the zero-sum game by building a swimming pool and leaving the lakeside alone. Yet, the very effort to get beyond reification and polarization is central to the task of social scientific analysis.

Still another illustration of the special problem that social contradictions have for the planners occurs when geographic and demographic changes are involved. One recent report simply noted that

> Wherever mass migrations occur, the interests of two opposing parties must be considered, and constructively provided for; the cultural integrity of the incoming people and the territorial integrity of the proprietary group. The law and its institutions can provide a structure for appropriate compromise, but the negotiation and arbitration will have to be personal and particular in each case. Because modern society requires cross-cultural communication and cooperation on an *intellectual* basis, spaces and social mechanisms for this must be created.[43]

There is altogether too much concern for synthetic solutions. Within an American context, it is the absorption of social science within social planning; within a Soviet context, it is the absorption of social planning within social science. Neither represents a better solution, and both leave intact and unchecked the relative powers of the political apparatus. In this sense, social science has the unique advantage of providing a critical role—critical in the sense of criticism—and in this way it can counterbalance the positivist and constructivist limits of a social planning approach. Clearly, planners should learn from social scientists, and vice versa. But such a learning process should not be viewed as an absorption process.

A profession is not a science, and a science may at times resist and even reject professionalism. In a democratic society there are plural mechanisms for expressions of wants and desires. There are constitutional safeguards provided by a system of checks and balances. No one really expects the legislative and executive branches to be identical, even though executives are chosen from the legislative branch, and executives may go into legislative activity. The same sorts of interaction might be encouraged in the areas of planning and social science: first, that systems of checks and balances rather than mutual commissions be established; and second, that there be arenas and forums of interaction and discussion without necessarily liquidating one discipline at the expense of the other through processes of either destruction or cooptation.

A revolution of falling expectations, or at least a sense of proportion, is coming to dominate American thinking: the bureaucratic legislation of the planning decades of the past quarter-century has come to an end. The work *The Urban Predicament*, published in the late seventies by Gorham and Glazer, is indicative of this new sense of modesty, forced in part, upon the planning community by the social scientists. The word "planning" itself is used even more cautiously. The authors indicate:

> We know that many things will not work: a simple expansion of expenditure, under which the greater share must inevitably go to the professionals in the fields of employment, crime, education, and housing, may offer little improvement in the quality of life or opportunity in declining neighborhoods; the creation of new community organizations through the infusion of outside funds seems to hold little promise at this point in history; the imposition of a *great plan* from the outside cannot, it seems, be responsive to the complex, interlinked problems that are dragging these neighborhoods down.[44]

Social scientists have noted that planning has its limits within a democratic context. For example, it might not be reasonable or even worthy to have everyone use public transportation. There may be aspects of private automobile travel so superior that all one can really do is to assume a basic continuing dependency on the private car, but then work toward the redesign of that private car. Hence, in the world of transportation, new voices are heard about "signals from the marketplace," and a consensus about what people are willing to adjust to develops, while policies designed to minimize political risk for legislators are advocated. The policy impact on planning in a wide variety of areas has been genuine, but limited.

The field of transportation has seen a big push in a rational model for public transportation facilities, especially to enhance the restoration of inner cities. But when it comes to circumscribing private vehicular transportation, planning agencies fail miserably to influence social behavior. In the area of public education, a wide array of alternative approaches, from a voucher system to busing students, have been introduced. But the results of policy efforts to institutionalize involuntary desegregation have been mixed. Any sense of success has been dampened by the fact that high levels of racial integration have been achieved through the unanticipated consequence of the outward flow from urban centers and a high increase in private school systems. This only serves to indicate that excessive planning, based on formal doctrines of policy-induced equity, may elicit a considerable backlash among the citizenry, including the presumed victims of inherited inequality.

In the area of crime, social scientists are less likely to argue the case for bigger and better police forces, but would instead direct attention to methodological problems and statistical estimates, pointing out that differences in crime rates observed across jurisdictions may not represent differences in criminal behavior as much as differences in the proportions of victims who report crimes to the police, or differences in the methods and skills with which local police departments report and record crimes. Again, the role of social science is clearly not to limit crime fighting but to indicate the kinds of issues that might be looked at more carefully than has been the case in the past.

It may well be that social science will become a repository for a kind of wisdom that is in direct conflict with social planning. Social planners, for their part, may be charged with defining areas of society and culture best left unplanned, unlegislated, and where simple market mechanisms or personal social preferences should be held as final arbiter of what is and what is not to be done. But whatever the specific relationships between these two groups, clearly the groups are not identical.

Let me conclude with a statement made by an outstanding representative of the democratic temper a quarter-century ago, Karl Mannheim. His warning on the gravity of the risks involved in separating planning from its imagined public still claims our attention; more so, given the rush of new nations to seek relief in the planning mode.

> Our task is to build a social system by planning, but planning of a special kind: it must be planning for freedom, subjected to democratic control; planning, but not restrictionist so as to favor group monopolies either of entrepreneurs or workers' associations, but "planning for plenty," i.e., full employment and

> full exploitation of resources; planning for social justice rather than absolute equality, with differentiation of rewards and status on the basis of genuine equality rather than privilege; planning not for a classless society but for one that abolishes the extremes of wealth and poverty; planning for cultural standards without "leveling down"—a planned transition making for progress without discarding what is valuable in tradition; planning that counteracts the dangers of a mass society by coordination of the means of social control but interfering only in cases of institutional or moral deterioration defined by collective criteria; planning for balance between centralization and dispersion of power; planning for gradual transformation of society in order to encourage the growth of personality; in short, planning but not regimentation.[45]

In the rush to professionalism, planners who ignore such social science warnings would do so with impunity. Given the fact that professional planning identity has been largely achieved in the absence of a critical social science standpoint, it becomes even less likely that the warnings and premonitions introduced by Mannheim will be taken as seriously as they should be. But until the basis for rapprochement is worked out between social planning, political mandate, and social science, let those differences that exist continue to form the basis for intellectual discourse and practical checks and balances.

4

Zero-Growth Economics and Egalitarian Politics

With admirable clarity, Lester Thurow has stated the problem of overdevelopment. He notes that, while most of our economic problems are capable of solution, indeed afford multiple solutions, "all these solutions have the characteristic that someone must suffer large economic losses. No one wants to volunteer for this role, and we have a political process that is incapable of forcing anyone to shoulder this burden. Everyone wants someone else to suffer the necessary economic losses, and as a consequence none of the possible solutions can be adopted."[1] A less value-laden way of stating this same proposition is that advanced societies have competing values driving them forward: egalitarian desires for the masses and elitist aims for the classes. To force collective social goals at the expense of economic innovation may result in a dangerous miscalculation of ends that would occur if one concentrated solely upon economic growth at the expense of social aims. It is the purpose of this chapter to help break this logjam by moving beyond artificial concepts of growth limits; for to do otherwise, to accept such limits to growth, is indeed to end up in a zero-sum game in which advanced societies become captive to their own dialectical formulas and mythologies.

The Limits to Growth debate initiated in the seventies by the Club of Rome has turned into a free-for-all on the limits to equity.

Some have concluded in no uncertain terms that "without rather radical changes in the consumption patterns in the rich countries, any pious talk about a new world economic order is humbug."[2] Others have coupled discussion about changes in consumption patterns with talk about the finite nature of physical and productive capabilities.[3] Meanwhile Third World demands for a radical redistribution of wealth are increasing. The debate has contributed the possibility of a new equation: the more limited the level of production, the more insistent are demands for a redistribution of present wealth.

In the United States, the debate is played out to its logical conclusion. Special interests have marshaled data to support their position—a standard lobby technique. Have-not groups have developed a tradition of this kind that is less venerable but no less compelling. Outsider groups have the benefit of social science forces which, if anything, are more sophisticated about collating data for their clientele than older lobbies, making the impact of their demands even greater.[4] As long as demands for a greater share of the metaphorical "pie" take place in a context of zero or limited growth, such interest-group and regional demands on the larger society are potentially volatile. In the past, American society met incremental demands by outsider groups such as blacks, women, youth, and ethnic groups by expanding productivity and consumption. Such demands are continuing to accelerate while productivity is stable or declining and consumption is curbed by environmental interests. Some examples of this phenomenon are in order.

First, there is the contest between the snowbelt and the sunbelt—between cities and states of the Midwest and Northeast, which are experiencing population stagnation or decline with no corresponding decline in demands for services, and those of the South and Southwest. Citing federal tax and spending policies which are shifting an enormous flow of wealth from the Midwest and Northeast to the South and Southwest, snowbelt states have attempted to shift federal tax money away from the sunbelt states, with some success.

Small town and rural people are also asserting demands for their share, arguing that the 31 percent of the nation's rural population includes 44 percent of the poor and 60 percent of substandard housing, yet receives only 27 percent of federal outlays for welfare and poverty. Critics say that rural America contains 29 percent of the national labor force but receives only 17 percent of federal expenditures for employment and manpower training programs. The complaints and demands for redress of grievances are generally accurate.[5]

Special-interest centers of all kinds cite facts and figures to support their positions. For example, the Center for American Women and Politics argues that women constitute more than 50 percent of the voting population and 40 percent of the American labor force but hold only 4 to 7 percent of all public offices. They note that few women serve in executive Cabinet posts; no woman has ever been appointed to the U.S. Supreme Court; and very few women have been governors or senators. Patterns of discrimination, the Center claims, hold at the level of federal, city, and state commissions as well.[6] Appeals for equity continue to be viewed as justifiable on statistical grounds.

There is a clear pattern of accelerating demand for equity at both national and international levels. America, despite its faltering economic base, is still viewed as showing the face of the future to other industrializing nations. But at the very time new special interests have emerged, demands for zero or limited industrial and demographic growth are being introduced. The result is an intense struggle for a redistribution of the American pie; for as the nation slows its growth and no longer tries to manufacture a larger pie, concern inevitably shifts to the size of the existing slices.

Limits to Growth or Limits to Equity?

Club of Rome recommendations clearly indicate that the sociological message concerning the limits to equity is the underpinning of the technological rhetoric about limits to growth. This clearly American initiative came forth with a series of statements that focused First World debate.

1. The Club's image of a "fixed pie" assumes resources for further growth to be non-renewable or limited. As a result there are few mechanisms available to prevent the rich from getting richer or the poor from becoming even poorer.

2. The Club foresees diminishing returns in that new tech-

nology and additional capital investment necessary to extract marginal resources will vastly increase pollution and exhaust resources. Diminishing marginal returns require an effort toward zero growth to re-establish the "natural balance."

3. The Club assumes that rapid change increases the complexity of problems, and hence makes resolving conflicts difficult and management of resources impossible. The Club recommends simplifying issues and centralizing decision making even if that requires limiting democracy.

4. The Club also assumes an uncontrolled expansion of population, which they see doubling every thirty years. Population growth will spur exhaustion of resources, and result in the inability to cope with the distribution of goods and services. Again, the Club of Rome seems to argue the case for stopping developmental impulses at a moment in time rather than fundamental redistribution of the gross world product.

5. The Club strongly hints that although progress may postpone the need for immediate drastic action, the final collapse would only be hastened and be even more severe. Again, there is the demand for restraint, even cutback.

6. Income gaps are widening, a circumstance that could bring about worldwide class war; political warfare seems imminent. The Club postulates freezing the current situation, which means ignoring current inequities, rather than urging redistribution of wealth.

The Club of Rome's recommendations range from controlled growth, to freezing the current situation, to, finally, the implementation of harsh, even repressive, measures to prevent possibly disastrous situations later in the century. At first, social science involvement in the debate concerned whether the energy crisis was a reality or a fabrication. Then came a realization that from real shortages sociological issues emerged and, beyond that, ideological issues. The environmentalists were also slow to take into account the economic costs of the protectionist policies they support.[7]

Special interest groups tended to discount the energy crisis. Leon Keyserling noted in 1975 that shortages of oil and higher costs of electricity resulted from economic scarcity, tight money, and rising interest rates that were particularly harsh medicine for public utilities.[8] There were some interest groups that failed to connect limits to growth with limits to equity, and simply saw the problem as temporary, the result of governmental and industrial policy making and therefore susceptible to influence and change.

From a political viewpoint, old coalitions have fragmented at the mass level no less than at the intellectual level. From the perspective of factory workers and factory owners, the environmental coalition is seen as a veritable conspiracy to prevent the exercise of free enterprise, free markets, and free labor. The relative reluctance of blacks and other working-class minorities to support the environmental-protection groups indicates just how differently they see their interests. An entire folklore has grown up that blames environmentalists, Arabs, and the oil companies for higher costs of operating and maintaining automobiles and for higher home heating costs. These costs are more easily borne by supporters of environmental interest groups, mainly the middle class, than by economically hard-pressed groups, mainly working-class.

Environmental groups have labored long and hard to persuade the public that environmentalism means new jobs, that the higher costs are relatively short term and worthwhile, particularly in poor and slum neighborhoods. Beyond that, environmentalists argue that they are drawing attention to oil and fuel shortages, so that solutions may be found to prevent the complete breakdown of the social system in the near future.

The Development of Equity Concepts in the Postwar World

However fervent the sociological debates of the 1950s about the tragedy of human inequality, there was a near consensus that stratification was a permanent feature of all economies. Hence the issue was not establishing egalitarian principles so much as lessening the extremes of inequality.[9]

The 1960s witnessed a nearly unanimous repudiation of this position. It was argued that social equality is both realizable and a practical necessity. Movements for equity and justice employed an array of statistics showing the existence of income and occupational gaps between the larger society and racial, sexual, and religious minorities. This was followed by open pleas for equity and veiled threats that this could entail parity of wealth, goods, and services.[10] The 1970s neither abandoned the goal of absolute equality nor reverted to an earlier imagery of the inevitability of inequality. A clearer distinction has emerged between what is and what ought to be, between the facts of inequality and the goals of equality. There has been intensified appreciation that equality is costly and, like growth, is subject to constant refinement. For if equality can be guaranteed only by growth, and growth can be secured only by differential rewards and specialized incentives (at least within market economies), then the problems of stratification and equality are dialectically intermixed. This fact itself may contribute to the current inability to forge new policies and anticipate new contingencies.

The eighties and possibly beyond are tending toward universal acceptance of equality as a goal and the parallel lack of acceptance of inequality. Earlier principles of self-regulation and self-interest have clearly fallen on hard times. Even those who maintain their fervent belief in individualism as a way of life admit the need for state regulation and social interests. The practical problems, which are often seen as resolvable by law, invariably concern *how much* regulation, *how much* supervision. Fred Hirsch has described this peculiar dialectic well:

> The principle of self-interest is incomplete as a social organizing device. It operates effectively only in tandem with some supporting social principle. While the need for modifications in laissez-faire in public policies has been increasingly accepted, the need for qualifications to self-interested behavior by individuals has been increasingly neglected. Yet correctives to laissez-faire increase rather than decrease reliance on some degree of social responsibility in individual behavior. The attempt has been made to erect an increasingly explicit social organization without a supporting social morality. The result has been a structural strain on both the market mechanism and the political mechanism designed to regulate and supplement it. In this way, the foundations of the market system have been weakened, while its general behavioral norm of acting on the criterion of self-interest has won ever-widening acceptance.[11]

The social system generates strong contradictions without providing mechanisms for resolution of ensuing conflict. Arguing that "small is beautiful" and urging limits to growth have the overriding effect of reducing tensions without necessarily altering structures.

How does the new egalitarianism confront the new conservatism? Herbert Gans emphasizes that the black movement provided a model for equality demands, and he presents three major bases of the "new egalitarianism":

> First, many American are now beginning to realize that the frontier, by which I mean the opportunity to strike out on one's own and perhaps to strike it rich, is closing down. Second, as people have voiced more political demands, they have also become less patient with political inequality, particularly with their increasing powerlessness as bureaucracies and corporations continue to get bigger. Third, the affluence of the post-World War II era has enabled many Americans to raise their incomes to a point where they are no longer occupied solely with making ends meet. As a result, new expectations have emerged, not only for a higher standard of living but also for improvements in the quality of life and for greater power to control one's destiny.[12]

These three points actually represent precipitating factors made possible by underlying structural changes. The factors require further elaboration, but first let us review the underlying structural changes that have brought them about and given them force.

The foremost structural change is the size, role, and concentration of state power in America. The federal government has become a large, complex organization, and most of its growth has taken place in this century. A first goal of the federal government was to provide an environment for business activity, with the creation of a stable currency and a stable legal order to ensure the rights of exchange and contract the first steps toward this goal. Besides this basic framework for business activity, government also provided support services and facilities which individuals and firms could not provide for themselves. For example, national roadways and waterways assist commerce. Indirect subsidies, such as protective tariffs and import quotas, and direct subsidies, such as those for the railroads, support economic activity. This legal-rational government provides a predictable environment for the growth of business and capitalist activity. Government legitimates itself by presenting its activities as favorable to economic growth and industrial expansion. As Calvin Coolidge said, "The business of government is business." Even reform movements such as that of the Progressives sought to use the state to provide a better business atmosphere. Anti-trust legislation was intended to check the evils of monopoly in business.

With the Depression and the New Deal, the role of government began to change. Government sought not only to improve business conditions but to better the condition of the population as a whole. Government not only created a framework for business activity which could benefit individuals through their upward mobility but attempted to intervene directly in the economy to uplift entire groups of people. Labor unions were recognized, and laws were passed favorable to their activities. Social welfare programs grew, both during and after the New Deal. By these actions the state indicated that inequality did not necessarily result from an individual's failure to take advantage of opportunities but possibly from social factors beyond the control of individuals. The state became the protector of the business community, but also took responsibility for improving the status of all groups. Presenting its goals in these terms, the state further delegitimated existing inequalities by pronouncing them not natural and therefore correctable.

Rossi has described movement from the lowest "pluralist" level to a higher "hybrid" model, as equality demands increase. Most equality movements create rising expectations. Rossi shows the effects of such ideologies, not only those of women but also those of racial and ethnic minorities.

> *Pluralist model:* This model anticipates a society in which marked racial, religious, and ethnic differences are retained and valued for their diversity, yielding a heterogeneous society in which it is hoped cultural strength is increased by the diverse strands making up the whole society. *Assimilation model:* This model anticipates a society in which the minority groups are gradually absorbed into the mainstream by losing their distinguishing characteristics and acquiring the language, occupational skills, and life style of the majority of the host culture. *Hybrid model:* This model anticipates a society in which there

> is change in both the ascendant group and the minority group—a "melting pot" hybrid requiring changes not only in blacks and Jews and women, but white male Protestants as well.[13]

Early on, equality movements were directed against individual firms or groups or involved opposing classes in society. Unions directed their activities against corporations or businesses by means of strikes, organizing political campaigns, or product boycotts. At first unions feared and opposed government intervention in this conflict. Today the women's movement, the black movement, and other equality movements aim more of their activities at the state than at any other part of society. The women's movement is trying to achieve equality through legislation or judicial review, through lobbying and electoral politics. Action against firms is not taken directly, but through a third party, the government, by means of the courts or human rights commissions. When direct action is taken against a business firm, it is often done to provoke the state into action. The limiting of opportunities in the private sector does not create strains that lead to demands for equality, so much as the opening of the public sector as a new avenue for equality demands does. It is only when both sectors exhibit closure that demands for drastic socio-economic change become widespread.

S. M. Miller and Pamela Roby have observed that power is an important aspect of the drive for equality.

> In the "welfare state," in particular, many important elements of the command over resources become available as public services. The distribution and quality of these public services affect the absolute and relative well-being of all individuals. Considerable inconsistency may exist between the incomes and basic services of persons or groups. While the two are fairly closely linked in the United States, poor basic services are *not* associated with low income in Sweden. A larger issue is also involved. As Marshall has argued, the welfare state approach breaks the link between the market and well-being.[14]

As the state attempts to legitimate itself as a service organization, political power becomes a major factor.

Along with increased state power and new expectations regarding equality has come the activation of civic participation in the lower classes. More groups take part in state politics. At first, expansion of citizen participation was expected to introduce a new era of political democracy and consensus, an "end of ideology." Paradoxically, while a nonideological period may have prevailed for a limited time, it laid the foundation for a new ideological period.[15]

As the state became a mediator of class claims rather than simply the legal expression of the domination of one class, the state was changed from a superstructural reflex to a foundation initiator. The class state yielded to the service state. The resulting delegitimation of inequality set the stage for greater political struggle as the focus shifted from the workshop to the Capitol and the community. The democratic state became the intended source of relief from inequality resulting from economic exploitation.

Economic Dysfunctions and the Social Costs of Inequality

A central premise of sociology and economics has been that inequality is a functional necessity for any society. Even Marxism states that classes are necessary for capitalist development and hence a precondition for the development of socialism. In sociology, the basis of this argument was presented by Davis and Moore:

> The main functional necessity explaining the universal presence of stratification is precisely the requirement faced by any society of placing and motivating individuals in the social structure. As a functioning mechanism, a society must somehow distribute its members in social positions and induce them to perform the duties of these positions.[16]

Similarly, neo-classical economics argues that within a market system inequalities in rewards are necessary. Without inequalities, persons would lack motivation to incur the costs involved in leaving present positions and taking on socially important positions that are unoccupied. Unequal rewards are said to motivate economic growth by spurring investment and entrepreneurship and hence a better society. Neo-classical economists cite the cost of equality in terms of the decline of efficiency. Attempts to achieve equality, neo-classicists believe, result in a less efficient economy owing to a decline in motivation, a drop in investment, and the rise of costly bureaucracies to ensure equality.

Most arguments against inequality have assumed the moral or ethical values of equality or, in its Marxist form, the iron laws of history (the Marxist argument also carries with it a moral vision). Most of these arguments have cited the improverishment or deprived condition of the lower classes and speak of the right to happiness or to better life chances. Christopher Jencks earlier in his career made this type of argument: "Why, after all, should we be so concerned about economic equality? We begin with the premise that every individual's happiness is of equal value. From this, it is a short step to Bentham's dictum that society should be organized so as to provide the greatest good for the greatest number."[17]

Another argument against inequality avoids the pitfalls of moral dicta or assumptions about historical laws. It points out that while inequality *may* be functional, it is also dysfunctional, a proposition that holds for not only those who have unequal positions, but also for society as a whole. However, most arguments concerning inequality, whether pro or con, fail to cite the possible social as well as economic costs.

Despite the arguments by Davis, Moore, and others, inequality may create a lack of integration in a society. As Goldthorpe pointed out, ". . . the existence of inequality, of an extreme, unyielding, and largely illegitimate kind, does militate seriously against any stable normative regulation in the economic sphere—because it militates against the possibility of effective value consensus on the distribution of economic, and other resources and rewards."[18]

The malintegration associated with inequality is not simple Marxist class conflict; it is far more complex. There is no clearer example of this complexity than crime. Although the literature does not agree about what is meant by class or status, a standard proposition in criminology is that certain crime, particularly violent crime, is negatively correlated with class or status. The lower the class or status, the higher the incidence of violent crime. "A 1960 Milwaukee study indicated that the slum or inner core area of the city, comprising 13.7 percent of the population, had 38 percent of the arrests for burglary, had 69 percent of the aggravated assaults, 47 percent of other assaults, 60 percent of the murders, 72 percent of the arrests for commercial vice, 22 percent of the drunkenness, and 67 percent of the narcotics arrests."[19] Similarly, large cities have a high rate of violent crime. They are more likely to contain concentrations of the poor and minority groups, and high crime rates are an indicator of the relation of crime and inequality.

Differences between small and large cities cannot be explained solely by greater objective inequality in large cities. Goldthorpe speaks of illegitimate inequality which raises the concept of relative deprivation. Analyzing relative and absolute deprivation is more complicated than was at first thought Poverty in itself is rarely a cause of crime, whereas resentment of poverty in an affluent environment is more likely to develop criminal behavior among the relatively deprived. In this sense development itself may serve as a stimulant to deviance.[20] A sense of relative deprivation is more likely to form in large heterogeneous cities than in small homogeneous cities. In large cities, interaction between groups is more likely, allowing reference groups to form and a sense of the delegitimation of one's own lower status to be felt.

The cost of crime resulting from delegitimated inequality (and not from poverty alone) can be estimated by the cash value of damage done by crime and of government expenditures to combat or correct the damages of crime. These costs and their relation to inequality become clearer when one considers business location. Controlling for ghetto, non-ghetto (inner city), and suburbs, one can almost create a scale by level of inequality. The lowest level is ghetto, suffering from economic inequality and racial discrimination. Next is non-ghetto, with economic inequality yet a mostly white population. The last of these groups is suburban. While these are not necessarily Warner's upper-upper class, they are relatively more prosperous. As one moves up this scale, crime decreases in most categories. This lends more support to the proposition about the relation between inequality and instances of crime (Bureau of the Census, 1975, pp. 159–60). As inequality becomes delegitimated, its direct cost increases. Direct monetary costs of crime include federal outlays for crime reduction ($2,839 million in 1975) and state and local government police expenditures ($6,535 million in 1973).[21]

There are at least two other dysfunctional consequences of crime. One is opportunity costs of the expenditure of funds as a result of crime: the use of limited government funds in criminal justice programs and police activities prevents their use in other programs. The second consequence is of non-economic social disruption resulting from crime. Durkheim may be correct in surmising that some level of crime may be functional by virtue of its shock value and reinforcement of social norms, but the present level of crime is far higher than this. Existing crime inflicts costs in terms of fear among non-criminal members of society. These costs are hard to quantify, but may include the fear of leaving home at night, lost sleep from strange noises outside a window at night, and so on.

Another dysfunctional consequence of inequality is diminished economic production. The dysfunctional consequences of inequality in the workplace have been obvious since Marx wrote about alienation and class conflict. Although in the United States there is no class consciousness or large-scale class conflict as described by Marx, inequalities still impose costs on the production of goods and services. In the workplace, inequalities are not defined solely by income. Ely Chinoy's study of automobile workers[22] demonstrated that workers rank their positions by type of work done, physical demands of work, health and safety conditions of jobs, and authority relations on the job—all in addition to income. This multidimensional ranking of work sets up numerous sources of inequality. And since these inequalities were pointed out by workers, their legitimacy may be questioned.

The consequences of these types of inequality are obvious. One measure of this cost is labor time lost owing to labor disputes. Work stoppages result from causes ranging from general wage dissatisfaction to plant administration. They are some indication of the degree to which differential rewards lower economic output, quite apart from the "ethical" question of differential payment for different work done.

Other costs of inequality in work are harder to measure. There is the inefficiency and poor quality of work resulting from discontent among those producing it. An indirect measure of these costs is time lost from work because of illness. Such a measure cuts across social systems and into the marrow of reward syndromes. Illness is not disease. Kadushin notes that "disease is abnormal structure of functioning; illness is the feeling of discomfort which arises out of disease," and goes on to show that there is little or no association between social class and *disease*, yet a high correlation between social class and *illness*.[23] A plausible explanation of the greater absenteeism by lower income groups is discontent with their work. Costs of inequality may also include sabotage on assembly lines, general apathy about the quality of products manufactured, and a breakdown of creative performance as a result of the imposition of greater routine.

There is little empirical work concerning the impact of no-growth policies upon employment or of increased employment upon the environment. Environmentalists and those supporting increased employment live in different ideological realms and have little contact with each other. Environmental groups view problems in terms of biology and physics with little insight into the social consequences of their policies. Groups supporting full employment and increased equality seem to believe the world still has infinite resources.

Economists concerned about the environment describe the negative impact of growth upon "space-ship Earth" and show little interest or concern about the consequences of no-growth upon employment patterns and equity demands.[24] A much different viewpoint has been expressed by Bayard Rustin, who saw the country as being in a historic national crisis:

> There is mass unemployment, a sizeable and expanding black underclass comprising persons whose lives were scarcely touched by the civil-rights revolution, and a declining standard of living for millions of working people. Yet many of those who profess concern about unemployment and poverty also actively support the concept of limiting economic growth in order to protect the environment. That notion, if translated into conscious policy, would measurably worsen the nation's—and the world's—economic plight. And its promoters would bear the responsibility for having shattered the hopes of those who have never had a normal role in the world economy, among whom the darker-skinned people of the world rank most prominently.[25]

Clearly, the line between problems of production and those of consumption is thin. Perhaps it would be appropriate to place greater emphasis on patterns of distribution—of money, property, and goods—and less on the mechanical distinction between ownership and operation of the means of production. Not to do so means to freeze the economy at high rates of productivity, which has little significant bearing on the distribution of wealth.

Environmental protection policy has direct and indirect effects upon employment and equality. One direct effect is the closing down of an operation or industrial plant because of violation of pollution laws, which lowers employment and creates economic depression in the surrounding area. For example, the Environmental Protection Agency acted to stop an ore-processing plant from dumping chemical waste into Lake Superior. The cost of complying exceeded the profits of the operation, causing the corporation to close the plant, which depressed the local economy. Older industrial cities bear the brunt of environmental laws. They generally have the worst air pollution and water problems and hence experience the strongest enforcement of environmental legislation. These cities have also become centers for low-income and minority populations. Industries in these areas, when forced to add expensive pollution-control devices, move plants to areas where incremental increases in emissions have less effect upon air and water quality. This results in limited increases or actual decreases in employment. It should also be noted that the jobs lost are generally unskilled or semi-skilled entry-level positions.

The indirect effect takes place in the context of structural changes in the American economy. This shift, as outlined by Daniel Bell, involves moving from energy-intensive productive industries to knowledge-intensive service industries.[26] The implication is that no-growth in manufacturing, in contrast to accelerated growth in service industries, is a social fact. As Bell states, this process is already going on, and environmental laws will only hasten this change. This shift has two implications for employment and equality. The first concerns job mobility. High-level positions in service industries require skill and knowledge; entry into these positions requires certified education and professional standing. Such industries create a professional elite and a vast class of unskilled workers with little opportunity for professional advancement.

A further implication for equality is the wage structure and productivity of services. Service industries suffer from low productivity and lack means to increase that productivity. Since wage increases cannot be based on increases in productivity, they will be harder to achieve. This is particularly true when the employer is the government and the wages are drawn from taxes. The relative decline of labor unions has further implications. Labor unions historically have been a force working for equality. As the industrial sector declines, the relative power of unions will also decline.

Thus far we have discussed costs of inequality in a limited context, i.e. when existing inequalities are perceived as illegitimate. Both crime and work conflict are costs inflicted by those who perceive their inequality as unnatural or wrongly and unjustly imposed upon them. There are other costs resulting from inequality that do not require either that inequality be delegitimated or that those in unequal conditions take conscious action. These costs center on less than optimal distribution and use of resources for which there is inelastic demand. Such costs are of particular importance for a world of limited resources and slow or no economic growth. And these costs are highest in a market economy. Mark Kelman presents evidence for this in relation to medical services.[27] The demand for medical services is nearly unlimited, yet supply is limited; as a result, physicians and hospitals are located in higher income areas, so that lower income areas suffer from a deficit of services. A more direct cost to society is a general rise in the cost of medical services because of this pattern. In higher income areas, large sums of money are spent upon non-essential services, such as cosmetic surgery. This drain of resources and personnel raises the cost of essential and life-sustaining services for all users.

A more equitable distribution of funds for medical services would limit demand for non-essential services and increase the supply of resources and personnel for more important services. Growth without redistribution of wealth would only increase demand for medical services by all groups. It would create greater demand for non-essential services and drain off more resources from essential services. For this reason, the United States may have the most used (some claim the most over-used) medical system, and yet the most costly and inefficient delivery system. Kelman's model of the social costs of inequality can be applied to other limited resources with elastic demand, such as energy, food, and possibly environmental benefits. All are limited, yet higher income groups demand and can acquire a greater share. Limited supplies result in higher costs across the board.

For the most part, equity questions have been considered within a domestic context. Increasingly, some international economists and sociologists, among whom Gunnar Myrdal is typical, have argued that a more appropriate context for these issues is international, and that problems of redistribution of wealth can be understood only within that larger context:

> The blunt truth is that without rather radical changes in the consumption patterns in the rich countries, any pious talk about a new world economic order is humbug. It is legitimate for an economist to analyze the rational inferences in regard to economic policy based on what is in people's true interests and their acclaimed ideals. But if, instead, we raise the problems of what is actually going to happen, it is difficult to believe that rational policy conclusions will be followed in the practical policies of the developed countries. In the tradition of Western civilization we are quite well trained to combine base behavior with high ideals.[28]

Whether or not Myrdal's call for "rational national planning," which has as its end product a "curtailment of consumption," is feasible, the fact that such desperate theorizing is taking place is a clear indication of the re-emergence of Malthusian doctrine, and a far less sanguine attitude to the existence, much less fair distribution, of basic resources.

The current stratification of nations is being questioned: not only the emergence of a Fourth World, a lumpenproletariat of nations in which neither food nor raw materials are available in adequate supply, but major shifts in relations within and between the established three worlds. Relations based upon ideology are called into question, and relations of dependency are reversed. The major consequence of the neo-Malthusian world is growth of power based upon raw materials, and the decline of power based upon technology or military position. At the same time the role of ideology diminishes. It no longer functions as a basis of alliance or as a block to alliance.

In both of these areas, the Organization of Petroleum Exporting Countries (OPEC) can serve as a useful model. The nations that make up OPEC are Third World nations and have Third World social organization, yet they differ from one another in many other respects. Although we generally consider OPEC to be an Arab cartel, it contains non-Arab Third World countries such as Nigeria, Indonesia, and Venezuela. These nations represent not only diverse religions and cultures but also diverse ideological views. Among Arab members are radical Algeria and conservative Saudi Arabia. Norman Girvan has outlined the dynamics of this economic nationalism: "Nor is Third World economic nationalism seen as stopping at attempts to control prices for the primary products upon which the export incomes and the economic livelihood of these countries depend. Market power is seen as only one component in a general strategy for securing control over marketing and ultimately over production of the natural resources that sustain the Third World economies."[29] This type of ideology fits well with either a socialist or capitalist economy, or with the mixed market systems that fall between the two.

As this new non-ideological ideology in the Third World has emerged, similar ideological fusions in the First and Second Worlds have broken up. An illustration of this fact is the behavior of NATO and OECD countries in response to OPEC. These countries failed to arrive at any common program of action during or after the 1973 embargo. At the same time, the Second World, which despite its own needs, is also petroleum exporting, acted in a non-ideological manner. During the Arab oil embargo, the Soviet Union gave full vocal support to Arab actions and yet continued to export to target countries. In the year of the embargo, Soviet exports to Japan and the Netherlands doubled, and the cash value of those exports tripled. This represented a highly pragmatic non-ideological worldview on the part of the Second World.

More important than the breakdown of ideology is the shift in bases of power. Until recently, power was based upon either technology or military capability. Within this framework, the First and Second Worlds were in a superior position to the Third World. With unlimited raw materials, industrial nations were able to set the price for both raw materials and finished products. With increased awareness of limited resources, this relation has been altered. Before 1973, for example, many predicted that Japan would be the dominant power of the next century. Japan is heavily industrialized and technologically advanced, but must import fuel and food. Japan was severely threatened by the embargo and took a number of years to recover fully from its economic effects. While oil is in short supply, manufactured goods are made readily available to OPEC nations. However, as the position of OPEC nations has improved, other Third World nations lacking natural resources have declined. The price of both raw and finished imports has been forced upward, and their limited exports are unchanged. This means that the Fourth World has more members.

While entrance into the middle class may not be the exclusive touchstone of upward social mobility, it is clear that limits to new energy and food resources create the same pressures for redistribution of existing goods at the international level as exist nationally. Keyfitz points out:

> Price increases such as those of the Organization of Petroleum Exporting Countries can have little overall effect on the number of middle class people in the world (although they have some effect on whether the newly middle class will speak Spanish or Arabic or English). Who ultimately bears the burden of such price raises is not clear. Some of the burden is carried by poor countries that are not endowed with raw materials; when the repercussions have worked themselves out, India may find it has contributed a higher proportion of its income to Saudi Arabian opulence than the U.S. has. Certainly some U.S. fertilizer that would have gone to India before 1973 now goes to the Middle East; German chemical-plant investments are similarly diverted. The offsetting of oil price rises by French arms sales to Iran has everything to do with national power and little to do with the total distribution of poverty or even the national distribution. The main point is that only a small fraction of the world population is in the resource-rich areas.[30]

Added to the woes of the non-OPEC Third World is the problem of whether the OPEC model can be used by countries with other raw materials. Four poor countries—Chile, Peru, Zambia, and Zaire (Congo)—supply most of the world's exportable surplus of copper. Three others—Malaysia, Bolivia, and Thailand—account for 70 percent of all tin entering international trade channels. Cuba and New Caledonia have well over half of the world's known reserves of nickel. The main known reserves of cobalt are in Zaire, Cuba, New Caledonia, and parts of Asia. And Mexico and Peru, along with Australia, account for 60 percent of the exportable supply of lead.[31] Some of these nation groups—for example, coffee, copper, and bauxite exporters—are trying to form cartels and gain more control over the marketing of their raw materials.

At the same time, there is some question as to whether the achievements of OPEC can be repeated by other Third World countries. Petroleum is an unusual product in that demand is nearly inelastic and substitution is difficult. A Brookings Institution study sets forth conditions that must exist for a successful cartel:

> a) the group must control a sufficiently large share of world exports, world production, and, for mineral resources, world reserves; b) the price elasticity of demand for the commodity in question, including the cross-elasticity with possible substitutes, must be sufficiently low; c) the group itself must be sufficiently cohesive to prevent individual members from pursuing their own advantage through unilateral action in the market.[32]

These conditions are not universally agreed to but do present some considerations that must be dealt with in any frame of reference. Few resource cartels could meet these conditions. Most non-fuel resources have elastic demand curves and can be substituted for or done without. For example, coffee can be done without; its absence would have no major effect upon the lifestyle of the user. Copper is easily substituted for, again without ramifications for the user. In comparison, petroleum, for both economic and social reasons, cannot be easily substituted for by other energy sources.

It is of interest to the First World that advanced technology does not fit this mode. The First World's power until now has rested upon its technological ability and control. Hans Morgenthau has stated that technology does not fit the framework above in two areas: monopoly and inelasticity of demand. The United States does not have a monopoly on advanced industrial and scientific technology. The expertise and production capability for a wide range of advanced technology products exist in most of the Western European countries and Japan. Over time, the possibility of a target nation's procuring advanced technology from nations other than the United States is likely to increase. There are few examples of advanced technology that are both essential and unique (that is, for which no substitutes are feasible). The First World has not been "sufficiently cohesive to prevent individual members from pursuing their own advantage. . . ."[33] The recent mixed decisions on the Olympic boycott illustrate this plurality of decision making. Two examples of the failure of technological power are the American trade embargo of Cuba and the United Nations embargo of Rhodesia. In fact, arguments for using food as an economic weapon are symptomatic of the failure of technology and industry to form a base of power in a neo-Malthusian world.

A no-growth policy will bring major changes in international stratification for both First World and Third World countries. It implies a decreased standard of living for much of the First World, and increased standards for Third World countries lucky enough to be able to exploit cartels. Since the Third World differs from the Fourth World by having a concept of emergence, the possibility of advancement for those members of the Third World who cannot play the cartel game is restricted in a world of limited resources. Therefore the size of the Fourth World will be expanded by these former members of the Third World.

The presumed "crisis" in world capitalism is largely a fictitious extrapolation from American conditions; it results from a disproportionate utilization of resources and energy in the past by the United States. In the 1980s, more, not fewer nation-states are involved in advanced stages of capitalist development, or at least a mixed market-welfare system that can boast a sizable, even expanding, private sector. The limits-to-economic-growth model is a response to specifically American limitations and should be so perceived before we are reduced to the manufacture of windmills.

The subtle transformation of a limits-to-growth model into a limits-to-equity model is fraught with considerable danger that deserves to be addressed. First, in shifting the emphasis to economic equilibrium and away from social stratification, certain structural deformities in American society are hardened, with the attendant risk of class polarization and ultimately class warfare. Second, to speak of limits to equity as an absolute physical requirement invites the state to turn a deaf ear to the needs and aims of the less economically advantaged sectors of society, another invitation to polarization. Third, to deal with such heightened polarization by fixing limits to growth is to encourage ever-increasing repressive measures internally. Fourth, such internal repression would inevitably be brought about at the risk of growing isolation from the international community of nations. Fifth, the ultimate consequence of the limits-to-growth argument is an absolute decline in the power of the First World, particularly the United States, an acceptance not only of the redivision of the world's resource and energy base but a redistribution of profits so that new wealthy and exploiting classes could act with impunity against the First World. This in turn would encourage deepening internal fissures between haves and have-nots domestically.

The limits-to-equity model is simply a mechanistic response to real structural changes in the balance of forces in the international economy and technology. To accept uncritically a model that incorporates an acquiescent response to the international status quo, and to deny categorically the major impulses that have guided American society, as the prototype of First World power in the twentieth century, is to invite a solution to a generally understood problem that can only sharpen the inner tensions of society and make it more vulnerable to contradiction and collapse.

5

Three Tactics in the Theory of Development

The study of international development involves a number of overlapping constituencies. The field was pioneered first by historians under the rubric of dynastic change. Then came the economists, who formalized the study of social transformation in terms of large-scale systems. Next came the anthropologists, who conceptualized processes of development in relativistic "then and now" terms, but nonetheless gave development studies broad cultural dimensions that had been absent until the post-World War II period. The field was enriched by political scientists, whose profound understanding of the role of state power, policy making, and authority moved the study of international development beyond a matter of inexorable economic trends or social tendencies and into decisions arrived at in local, national, and international exchanges.

Sociologists have not so much enriched empirical studies as they have developed the theoretical dimensions of development. Empirical work, especially after World War II, involved attitude and motivational studies transplanted from American to overseas contexts, using scales and measures largely developed in social and clinical psychology. While these studies gave sociologists input into the analysis of international development, the transformation of these psychological studies into general theory has been the more important contribution of sociologists.

The empirical contributions of sociology to the field of international development have been substantial, but they have focused at the level of explanation. The classics of sociology, however, make broad theoretical jumps from the data. The explanation by Marx of stages of development in terms of class formation and reformation; the explanation by Durkheim of the movement from organic to contractual foundations for establishing social solidarity; and, of course, Weber's emphasis on the role of religion and culture in creating a set of values that permit sustained growth—all are examples of such classic theoretical statements. They provide the background to current attitudes and orientations about international development.

Lipset's entrepreneurial thesis,[1] in which a "deviant" class of entrepreneurs creates the foundation for new values permitting capitalism to form in Latin America, is a direct descendant of the Weber hypothesis. At the other end of the spectrum is Frank's dependency theory.[2] In this view, the underdevelopment of the Third World is explained by the overdevelopment of the First World. This approach has clear roots in Leninism and its theory of imperialism, if not directly in Marxism. More than any other discipline in the social sciences, sociology may have inherited a broad nineteenth-century tradition of social theory. To that degree there is an unbroken line of theorizing about the processes of social change and international development.

Sociological Orientations Toward Development

Sociologists have played a central role in the creation of three orientations toward the study of development: the modernization, developmental, and dependency theses. These schools are characterized by the personnel involved, publications produced, the places where the work is primarily located, and, above all, by the intellectual positions they take. This last characteristic, the general ideology of development, may best explain exactly what sociology has meant to the field of international development.

The modernization school has been discussed in Chapter 2.

Basically, scholars such as Hoselitz,[3] Lerner,[4] and Shils[5] hold that modernization is the central source of development; just as traditionalism is the main source of stagnation, or non-development. There are infinite variations on this theme; each involves an assumption that the historical transition is from tradition to modernity, and the motor force in this transition is the infusion of highly sophisticated mechanisms of communication and transportation. As these permeate, the values of the society shift, and demands for participation in modern world systems accelerate.

The modernization thesis is informed by Keynesian economics, the idea of mixed economies of public and private sectors bargaining for the most sophisticated mechanisms through the impulse toward modernized systems and behavior. Generally the modernization thesis sees the central change as a shift from a feudal, landed agrarian economy to an urban-based industrial economy. The policy consequences of this shift are bargaining between interest groups and the description of class in terms of leverage factors controlled by each group. Modernization is perceived as a way to achieve change without revolution or reaction. Much of the modernization literature assumes that the higher the degree of social mobility, physical mobility, or information diffusion, the greater the degree of democratization of the social system. There is a strong implication that modernity is the fundamental expression of democratization.

The model for modernization was the advanced Western powers, primarily those who won World War II and thus stabilized capitalist democracy. Modernization clearly reflected the values of the American century: aspirations to participate in a consumer-based economy, universal communication, and cultural relativism. It represented the export of democratic individuals, of achievement-oriented societies. Continued stratification gaps, class, race, ethnic, and sexual disparities, and military definitions of world realities were magically filtered out of consideration. The entrepreneurial approach was exported to nations where entrepreneurs exhibited little instinct for risk investment or technological innovation. A mechanistic restatement of class values took place, with little understanding of the exaggerations inherent in international systems created by powerful donor nations on behalf of recipient nations. Remarkably, the modernization school held unquestioned sway in American social science for a long time. Despite the fact that it was transparently based on a single set of political and economic experiences, the position remained impervious to a sense of inner strain and failed to respond to external intellectual pressures.

The widespread emergence of authoritarian, totalitarian, and military regimes in the Third World, no less than the military emphasis in East-West Cold War politics, led to a reconsideration of modernization theory. A younger generation of theorists, such as Huntington,[6] Furtado,[7] and Dumont[8]—labeled developmentalists (despite their obvious ideological disparities)—perceived a breakdown in the identity between modernity and freedom. For this group, the relative position of any state or society in an international framework, or different sectors, classes, and subclasses within a state and society, affects events more than values, instincts, and attitudes. In a sense, developmentalism simply represented a long overdue shift from an emphasis on structural factors rather than personality factors.

Historical evolution was no longer presented in terms of movement from traditional societies to modern societies, but rather from colonialism to independence. Emphasis on Europe and America gave way to an appreciation of Asia, Africa, and Latin America. The developmental process was understood to be triangular in character: with economic classes pre-eminent in the formation of capitalism, political classes central to the formation of socialism, and military classes pre-eminent in the evolution and development of the Third World. In the developmental framework, policy became an even more important aspect of development, since questions of capitalism or socialism, democracy or authoritarianism, were matters of choice for new nations rather than historical inevitabilities. At the same time, decisions concerning reform and/or revolution were determined by strategies and tactics, and not simply models of social change imported from an earlier European framework.

The model for developmentalism was Third World nations that had been successful in achieving independence after World War II. Nationalist revolutions amalgamated what they held to be the most feasible structural components of the First and Second Worlds. The developmentalist position took note of strong Third World/new nation political commitment to one-party rule, extensive military leveraging of the social order, and the emergence of a bureaucratic sector having a specific set of planning and developmental tasks. The Third World took its economic cues from Western society, specifically the Keynesian revolution in marketing, exchange, and circulation of capital. As a result, consumer values and the instinct for commodity acquisitions, far from being crushed in these Third World states, emerged to full flowering. The Third World, which first appeared on the scene as a strange eclectic amalgam of communist politics and capitalist economics, gradually evolved its own symmetries as well as its own structures. The developmentalist perspective responded to the stability and durability of this new situation; it rejected the idea that the Third World was simply a transitional phase marching toward modernization, or for that matter, marching toward socialism. The new nations were indeed marching, but to a beat set by their own leaders. The analysis of this new beat gave substance to the developmental frame of reference.

In more recent years, the dependency school has attempted to restore a sense of holism and hegemony to the study of development. Under the guidance of Wallerstein,[9] Baran,[10] and Johnson,[11] the emphasis has been placed on power theory rather than either value theory (modernization) or interest theory (developmentalism). The dependency school assumes that the state of backwardness of most Third World countries results not simply from variations in historical evolution but from the conscious manipulation by the advanced First World of a dominated Third World. The idea of imperialism became central for the dependency school, as had the idea of class for the developmental school and the idea of nation for the modernizing school.

The world of dependency theorists was not so much bifurcated between modern and traditional sectors, as in the modernization school, not so much tripartite (First, Second, and Third World), as in the developmental school, but rather a unified world system that could best be understood in terms of position in the core, semiperiphery, or periphery of power and dominion. The dependency group was less interested in policy making for the Third World than the need to make revolution a precondition for policy transformation anywhere in that world.

The dependency model clearly took its cue from Lenin's theory of imperialism, from the notion that cultural backwardness and economic deprivation resulted not merely from the histories of Third World countries but from the excessive power, in international terms, of advanced capitalist nations. If Leninism is Marxism in the era of imperialism, then it might be said that dependency theory is Leninism in the era of post-industrialism. Dependency theory was given practical impulse by the realities of adventurous American foreign policy in Latin America, involving the use of the military or surrogates in nations ranging from Guatemala, the Dominican Republic, Cuba, Brazil, and Nicaragua, to indirect participation in the management and manipulation of nearly every other regime in the Western hemisphere. Largely for this reason, the dependency school of sociology was clearly rooted in United States/Latin American relationships. Hemispheric politics may have provided the context, but the Vietnam War provided the trigger mechanism for a general reconsideration of sociological standpoints. The conflict in Southeast Asia provided a seemingly overwhelming illustration of United States intervention in the affairs of dependent and small nations.

The ascendance of the dependency school, buoyed by American "defeats" in Vietnam and Cuba, turned out to be short-lived. The dissolution of the dependency school came slowly. The United States defeat in Vietnam, the growing independence of Latin American nations, and the expanding economic influence of the Middle East OPEC nations were in effect defeats for the excesses of American capitalism. As a system, capitalism showed signs of being strengthened rather than weakened by new developments. Vietnam and Cambodia became proxies and client states for their larger masters, the Soviet Union and China respectively. Latin America developed characteristics of national independence, military rule, and capitalist entrepreneurship. The OPEC nations developed their own notions of profit sharing, and that meant investment in, rather than destruction of capitalist bastions of power. The globalization of capitalism changed the ratio of power within the capitalist system but left the system as such undisturbed. Some dependency advocates actually saw the Soviet Union falling under the sway of capitalist blandishments and multinational offerings. The dependency model provided a sensitizing agency to imbalances in world power. It displayed a dazzling theoretical paradigm for the analysis of such imbalances. Dependency theorists globalized the notion of equity; egalitarianism was made into a worldwide concern. It was understood that the rights of individuals and collectivities must be adhered to quite beyond the boundaries of any one given nation or economy.

Sociological Schools of Thought and Their Ideological Institutionalization

The degree of overlap among these schools of thought is almost as great as the sense of integration within each of them. Yet clear lines do exist. Leading figures in the modernization school include Bert Hoselitz, Alex Inkeles, Wilber Schramm, S. M. Lipset, Kingsley Davis, Edward Shils, and Daniel Lerner. Clearly, there is an age cohort involved; most of these people did their primary field research in the fifties, although most are still quite active and vigorous. *Economic Development and Cultural Change* has been the foremost journal of the modernizing school for the past quarter-century. There has been a tendency over time to mute the ideological assumption that modernization and Americanism are one. However, there remains a continuing and clear persuasion that national development must employ modernization in all its parts: capital-intensive industry, bureaucratic politics, and mass society. That means changes in attitudes no less than growth in industrial output. The modernization school, because it developed first, and because it bore the full authority of sociological tradition, is perhaps strongest at Ivy League institutions such as Harvard and Princeton as well as at Stanford and Chicago, which also contain important institutes for the study of world communism and European politics. This strong bias toward an American ideology fit directly into the modernizing thesis that often indicated overlapping orientations toward the developing areas. Involved, too, was a confrontation in ideology and policy with Soviet power in the Third World. The modernization school, sometimes consciously, other times covertly, became party to the concept of a world struggle perceived as between a democratic West and a totalitarian East. There is really no way to disengage these two aspects of modernization theory; sociological theory and political sentiment both underwrote the modernization thesis.

Developmentalism involved people such as Aníbal Pinto, Allan

Schnaiberg, Charles Moskos, Denis Goulet, Gino Germani, Pablo González Casanova, Fernando Henrique Cardoso, and Raúl Prebisch. It was much more a phenomenon of the 1960s than of the 1950s. No doubt, it in part reflected a thaw in the Cold War, a termination of the Dulles era in the United States and the Stalin era in the Soviet Union. But it also included a much higher level of participation by scholars from the Third World, especially the advanced countries of Latin America. It, too, developed its own publication framework, primarily around *Studies in Comparative International Development,* which began as a monograph series in 1964 and evolved into a journal, and *Comparative Politics* in 1969. Institutions such as the University of Michigan, Northwestern, the University of California at Los Angeles, the Economic Commission for Latin America, Syracuse, the State University of New York, and the University of Pittsburgh, which came into prominence in the 1960s as graduate education in sociology rapidly expanded, became centers for the developmental approach in the United States. The developmental model also implied increased sensitivity to problems of policy and the bargaining aspects of power relations: seeing the Third World as presented with a series of options and decisions, rather than inexorable tendencies from tradition to modernization or, as was later to be the case with the dependency model, from feudalism to socialism. The developmental school also penetrated the structural variables in each of these societies, emphasizing how the processes of electrification and energy allocation shaped the behavior and demands of masses for migration and participation.

Just as the fifties gave rise to the modernization school, and the sixties to developmentalism, the re-emergence of Marxism as a respectable academic framework in the seventies gave impetus to dependency theory. Dependency theorists believed that the problem of international development could be best understood and examined in the bowels of multinational corporations and Western capitals rather than in underdeveloped peripheral societies. A new periodical issued by the Fernand Braudel Center (*Review*) is probably the best exemplar of this tendency, with *Perspectives on Latin America, Insurgent Sociologist,* and *Kapitalistate* representing a similar, albeit more strident point of view. Again, those who carry the torch in this school tend to be younger people, led perhaps by Immanuel Wallerstein, and including Richard Rubinson, Terrence Hopkins, Gabriel Kolko, Dale Johnson, André Gunder Frank, Susanne J. Bodenheimer, among others.

A problem in dependence theory is that it leads to analysis of dependency only as a United States phenomenon, with scant attention to the Soviet Union. Hence, dependency is seen as economic rather than political in character. Other problems have also arisen. The dependency school sometimes appears as the opposite of the modernization school, emphasizing unifying global factors, but with a strong critical posture toward the United States rather than the earlier critical emphasis on the USSR. There is also an ideological substitution of public enterprise for private enterprise as a mechanism for solving organizational problems, without much attention to problems of incentive, corruption, and innovation which are commonplace problems of socialist economies. Such problems remain to be worked out.

The Sociology of Development and the Development of Sociology

Ultimately one is left with a serious problem in the sociology of knowledge, no less than empirical analysis as such. Are we dealing with a sequence of intellectual positions moving from modernization to developmentalism to dependency, as one moves from the fifties, to the sixties, and to the seventies? Or can we evaluate these positions as alternative strategies in the pursuit of scientific prediction and explanation? If sociological participation in developmental studies is simply subject to a genetic explanation, all we could possibly hope for is a more interesting theory for the 1980s and 1990s. I suspect that the three postwar decades have outlined three alternatives, each of which will be tested in the crucible of events in the years ahead.

It might be the case that each position will experience modifications made imperative by certain methodological refinements. On the other hand, perhaps one of these three positions will triumph intellectually because it better explains events. That which comes first need not be last; nor will that which comes last in terms of evolution necessarily be intellectually triumphant.

There are limits of interpretation and integration in the field of development studies. Even if some rough correlations are possible between conservatism and modernization, liberalism and developmentalism, and radicalism and dependency theory, as some have argued, we are still left with a scientific decision about the most satisfactory and efficacious mode of analysis. By the same token, if we shift our orientation to satisfy contemporary fashion, that is, whatever particular paradigm is most prevalent among specialists at any given period in time, we must still confront the matter of truth, or the relationship of reality and experience.

My own confirmed belief is that the developmental thesis represents the mainstream of sociological good sense, and alone offers the possibility of a Third World perspective on itself and for itself. Developmentalism uniquely accounts for external pressures and internal dynamics in the growth process. It combines the best and most advanced techniques of qualitative and quantitative research procedures. It insists upon exact attention to ethnographic, linguistic, and national characteristics of those peoples and processes under investigation. Developmentalism does leave open the question of the fundamental construction of social reality; in this it is like the best of sociology generally. It makes the fewest a priori assumptions as to whether economic, political, social, or military factors are central in terms of their explanatory power. Developmentalism also leaves open the possibility that priorities may shift within a nation. Policy making, evaluation studies, and reshuffling social indicators may change the very structure of the development process; and developmentalism is uniquely situated to make such adjustments. The risks of eclecticism have been duly noted by the opponents of the developmental perspective. Yet, what are the alternatives?

Modernization offers a furtive "model of models," in which the measurement of development becomes the productive techniques of advanced industrial countries, their commodity fetishes, and the behavior and attitudes of the citizenry of these non-randomly selected nations. Hence, the measurement of modernization becomes inescapably linked to a celebration of Western capitalism. Such a viewpoint is extrinsic, and often alien, to the self-discovered needs of the developing areas themselves. Dependency theory offers a reverse side of this model of models in which the measurement of development becomes the productive techniques of advanced socialist countries, the social organization of these societies, and the behavior and attitudes of leaders of these countries. Moving beyond dependency becomes inexorably linked to participation in, if not the celebration of, Soviet communism. Such a viewpoint is extrinsic and alien to the grounded growth needs of developing areas. Developmentalism as a standpoint and an ideology offers the closest approximation to the structure of the social scientific community and to the needs of the Third World for an autonomic standpoint for assessing its own achievements and limitations. Developmentalism is the only theory that does not presume the Third World is somewhere in limbo—magically and mysteriously "on the road" to either modernization or socialization. Indeed, "the road" for the most part eludes analysis based on ideological assumptions. Developmentalism as a perspective takes seriously the contours of the national and regional requirements of developing regions as such. For these reasons, which I hope are objective, I argue that developmentalism has emerged as the master paradigm of social change in the twentieth century: the position that best satisfies the needs of social science research and political policy goals.

There is a significant but quite subjective distinction between modernization *and* dependency theories on one side and development theory on the other—the level of intensity brought to bear on the goals of a society. Modernization advocates, armed with functionalism, just *know* that the United States and the OECD nations represent the model of models and the future of futures. Dependency advocates, armed with historicism, know with equal or greater fervor that American imperialism is the source of all evil in the developmental process. They are perhaps somewhat less certain about which forms of socialism provide the best cure, but they are not above advocating a course of action accelerating the breakup of capitalism through revolution. The developmentalists lack any synthetic apriority and are reduced to uncertainty and intellectual hedging, which hardly makes for dedicated acolytes or clear-eyed emancipators. The situation might be characterized by Yeats's lines that "the best lack all conviction, while the worst are full of passionate intensity." In the world of social science, insofar as the science part is taken seriously, the monism of the hedgehog is a dubious distinction, while the pluralism of the fox has its own merits. Modernists and dependists, by disallowing the present to fulfill their skewed vision of the future, inevitably lead the independent researcher to inquire about teleological agendas and private motives. The developmentalist perspective has not been so clearly etched on the sociological conscience as have those of its adversaries; not because of confusion or eclecticism, but because it requires that social science be performed as an act of a courage born of direct investigation, rather than an act of faith born of presumed divination.

II

Militarization in Third World Development

6

From Dependency to Determinism in the Development Process

There exists a major discrepancy between inherited theories and established facts with respect to militarism in the Third World. I shall illustrate this by special reference to the case of Latin America. In the 1960s the United States provided 35 to 40 percent of all the armaments sold to Latin America and financed many additional purchases with dollar grants. In the 1970s the United States provided only 14 percent of Latin American arms purchases, or roughly one-third of what it provided in the 1960s.[1] Theories of dependency, of the consolidation of American hegemony, would lead us to anticipate a sharp reduction in the militarization of Latin America. Yet since 1945 not only has the military might grown, not only have arms purchases increased, but more regimes in Latin America are under direct military tutelage—often against American wishes—than ever before in this century.

There is a contradiction between dependency theory and social facts. This chapter addresses some of the questions this contradiction raises. The chapter is divided into three parts: first, a statement on the general character of militarism in present-day Latin America; second a somewhat longer section on what happens to modernization theories if we examine hemispheric affairs by holding constant the military factor, rotating, in qualitative terms at least, the economic, political, and social processes respectively; and third, the place of developmental versus dependency interpretations of hemispheric militarism. Understanding how and why the military has become central to Latin American development is a key objective. As Janowitz appropriately indicates: "The underlying question is not the conditions under which the military will 'exit from power' but rather how the long and twisted process of the transformation of the military regimes will take place."[2]

Military Structure of Latin America

Analysis of Latin American militarism has been inhibited by the tendency, inherited from older intellectual traditions, to perceive military intervention into politics as a function of class. Militarism is said to be a concomitant of oligarchical regimes;[3] part and parcel of middle-class pressures;[4] and, in regard to the Cuban Revolution, guardian of proletarian interests.[5] Without denying the existence of linkages between the military and social class, one can persuasively argue that the weakness or absence of class cohesion is a cause of direct military intervention. At any rate, at this stage of Latin American evolution the military must be taken seriously as an autonomous and independent agency of national power.

The choice of the limiting phrase "national power" is to emphasize the continuing boundaries of sovereignty. The economy, polity, military, and society are the four major pillars of modern national systems; a challenging task is to determine how these national systems interact, in bilateral as well as multilateral terms. The structured framework provided by sovereignty implies the central role of national development and the place of the economy, polity, society, and military in the definition and delineation of the nation. If the nation-state is an anachronism, it still informs and limits world systems of economy and power.

Latin American militarism is not accidental, nor is it a result of common geography and common hatreds. It is entirely a systemic phenomenon, predicated on the centrality of armed force and a corresponding lack of cohesion among socioeconomic classes and political organizations. When no class can exercise complete political legitimacy, the military, whether by design or default, assumes the tasks of governance. Charles Moskos, Jr., recently put the matter succinctly:

> This Western experience—via the military factor—has now come to encompass the world as well. A look around the globe reveals striking parallelisms in formal military structures, the transnational spread of advanced weapons systems, the incongruity of the poor nations of the Third World entering a period of military plenty, and the presumption among military professionals that they can speak the same "language" with each other. Even such seeming antipodes as United Nations peacekeeping forces and international terrorist groups are part and parcel of the globalization of the military factor. Such a conception fits as neatly on the contemporary scene as the *Homo economicus* model of a single world-economy.[6]

tarism is both a mechanism for and a consequence of social change

The importance of the military does not derive exclusively from the need to fill a class vacuum, nor is economic growth or political stability a necessary consequence of militarization. Mili-
in Latin America. It creates an administrative-bureaucratic cadre in the context of class stagnation or political vacuum.

Military regimes are neither more nor less stable than civilian regimes.[7] On major variables, military systems do not exhibit significant differences from non-military systems. Some military regimes produce a great GNP; others do not. Repeated military coups have tended to diminish rather than enhance the real rate of growth of exports. There is little evidence that systems of rule dominated by the armed forces spend more on military hardware than do comparable civilian regimes. In short, there is scant evidence that the militarization of Latin America is a cure-all for economic inflation or a counterforce to political instability. The breakdown of conventional explanations, coupled with the constancy of military dominion, indicates that the militarization of Latin America requires little extrinsic rationalization. Military rule is undertaken not to satisfy economic, political, or social claims but, more aptly, to satisfy the structural conditions of twentieth-century nationalism.

In the capitalist revolutions that began the industrialization process, economics clearly dominated all other elements of society. Long before the Russian liberation of the serfs of 1863, the French Revolution of 1789, or the American Revolution of 1776, the bourgeoisie had established its commercial supremacy. Political form followed economic fact. The marketplace dominated all other facets of Western civilization. The complaints of those who believed in a powerful state, of those who thought the bourgeoisie acted with disregard for the nation and its needs, were at least as great as if not greater than those of revolutionary sectors who believed that the bourgeoisie systematically ignored the needs of the poor and the downtrodden. The First World displaced medieval castles with the marketplace and stock exchange before appropriate forms of state power emerged which corresponded to the class needs of the bourgeoisie. As a result, what we call the "First World" was defined by its economic contours. The monied classes ruled the world of farm and factory alike. The political element was an afterthought, a necessary scaffold to support bourgeois power; the military factor seemed even more remote in forging the great revolutions of the West. Military conflict was real enough; but wars invariably were a response to the need for independence, so the military was viewed as armed professional adjuncts to civilian searches for economic liberty. This was certainly true of the American Revolution, where the "unbroken span of effective control of the military services by civilian authority is one of the most striking facets of American history."[8]

In the socialist world, certainly as defined by its first major success, the Soviet Union, political form preceded economic change. Socialists argued for spontaneous economic growth (presumably the outcome of capitalism exhausting its structural capabilities), but the communist position attacked the theory of spontaneity and asserted the need for forthright action embodied in the dictatorship of the proletariat. Power was to be seized by controlling the state apparatus; consolidated by legalizing the proletarian class and political aims; and legitimated by a new economic order based on prediction and planning. The political revolution occurred with the idea of socialism, not the fact of socialism, at hand. This is a direct reversal of the causal pattern in the capitalist world.

Like the First World, the Second World developed a military stratagem that was primarily intended to support and augment the secular political effort. The armed forces were not initially viewed as an autonomous arm of political rule, but rather as a professional adjunct to the political process. The Red Army was part of the larger political process. It came about during the civil war period, several years after the seizure of power by the Bolshevik party.[9] The Soviet Union offers a clear example of how the political process comes to dominate the entire society, especially its economic processes. Bureaucratic growth comes about through accession to state power; mobility is defined in political rather than economic terms; getting ahead takes on administrative rather than financial dimensions. Social stratification is real, even acute in the Soviet Union; but the access routes are political rather than economic. Trotsky and other dethroned leaders of the Bolshevik movement maintained that the political process was the soul and essence of socialism. In this sense, the determination by the state of all facets of economic and social life was built into the political system of socialism in the Second World.

In the Third World, Latin America is not only typical but prototypical, since it is the only sector of the Third World which has old nations intact and hence firm cultural patterns. The Third World rests on the military determination of power. The classes that made national independence possible were weak and unable to rule. The political system that came about after independence was too fragile to absorb the needs of all social classes and also too weak to control the bourgeoisie. As a result, the military came to power in nation after nation of Latin America (and later in sub-Saharan Africa and Southeast Asia under both direct and indirect colonization).

Charismatic dictators with loyal armed followings took control of the national armies. This political-military congruence provided the vanguard and mass base for Latin America's liberation from internal and external colonialism and also acted as a developmental stimulus. The military was not an adjunct to economic or political power as in the First and Second Worlds; it was an independent source of power. It delineated what form the economy would assume and what the established political system would tolerate. The spirit of military nationalism was for the most part anti-capitalist and anti-statist. On the other hand, the soul of the nationalist effort, the implementation mechanism behind the wave of anti-colonialism that drowned its opposition, was clearly the armed forces.

In Latin America, peculiar, or at least seemingly implausible, compromises have been registered. For the most part, the nations' economies remain firmly within the sphere of capitalist relations of production and free-market patterns of consumption. Even if these mechanisms are state managed, the so-called public sector is still very much involved in the private accumulation of capital on terms profitable for investors. At times this decision is made on consciously ideological grounds, i.e. neo-Keynesianism. In part as a consequence of the needs of militarism, and in part because there was no felt need for the political artifacts of bourgeois democracy, Latin America has accommodated itself to a Leninist model of politics in which elections are at best educational affairs and at worst do not occur at all. Above all, there is a close connection between the bureaucratic machinery and political apparatus. As Celso Furtado has pointed out, Latin America is a kind of shopping basket.[10] The basket contains capitalist economic realities and statist political systems. The shopper is the military. What is bought and what is not bought, what is retained and what is discarded, is essentially determined by the military. Any description of Latin American societies that does not begin with this simple recognition of military pre-eminence is analytically doomed. The first premise of any analysis of the Latin American political system must be this: militarism is not simply an accident of history or the curse of our times but is endemic and intrinsic to the structure of Latin American social relations.

The Military Factor in Latin America

The military in Latin America is pervasive and powerful. Political and economic systems as widely diverse as those of Brazil, Argentina, Haiti, Bolivia, and Cuba all boast powerful military machines. Each country displays variations and nuances, such as the relative weight of police vis-à-vis the army, or paramilitary units versus regular military units, but in all of them decision making takes place within the broad framework of military systems.

Even such political democracies as Mexico and Venezuela have not escaped the military influence characteristic of the area. Mexico has a single-party system with apparatuses rather than oppositions at the political level. The growth in Mexico's military sector has made PRI (Partido Revolucionario Institucional) beholden as well as responsive to the military. Venezuela, that bulwark of South

American democracy, shows signs of "Mexicanization," that is, it is developing a pattern of military growth in which the hardware components are real rather than symbolic. Older established democracies like Uruguay and Chile have become military systems with no apparent democratic goals, and elsewhere civilian regimes boast powerful military directorates.

The Latin American military believes itself to be centrist in character. Although from distant shores military regimes such as those in Brazil, Peru, and Argentina seem profoundly rightist or leftist in character and structure, these regimes define themselves as seeking options to organizational and ideological extremes. For instance, recent Brazilian regimes claim with justification interest in gradually restoring a parliamentary balance with a strong two-party system. In Brazil the military opposes and is opposed by rightist as well as leftist elements. Terrorist attacks by the Aliança Anticomunista Brasileira against the left-oriented Conferência Nacional dos Bispos do Brasil have had the unintended consequence of strengthening the centrist potential of the military.[11] The Peruvian military regime, while closer to leftist agrarian reform movements in origin than its Brazilian counterpart, likewise seeks to balance and correct ideological extremes. The former president, Francisco Morales Bermúdez, managed to eliminate both rightist Carlos Bobbio and leftist Jorge Fernández Maldonado in a "bloodless coup." The heated rhetoric of non-alignment that substituted words for deeds was also disposed of, as was the reform regime itself.[12] The technocratic elements within the Peruvian military are in charge, whether they wear nationalist tunics or multinationalist suits.[13]

The Argentine military considers itself in a fight to the finish with the left-wing Montoneros and right-wing Tacuaristas and their paramility equivalents—the terrorist forces of the Left and the police forces of the Right.[14] In Bolivia the military regime of General Hugo Banzer Suárez moved from a simple seizure of power to a fully articulated "philosophy" of military dominion. Even if there are limits to such military rule in countries like Mexico, Venezuela, Costa Rica, and Colombia, the overall thrust toward military determination of the political process is so powerful that it affects the very foundations of contemporary Latin American societies. Failure to understand the technocratic and anti-ideological nature of the military has led first to profound errors of analysis, and then to a view of the military as some dependent superstructural creature beholden to older class formations or overseas economic tendencies. The armed forces of Latin America, though interdependent, are autonomous. To deny this is to falsify the significance of military determinism in South America.

All too frequently, the military is seen either in narrow political terms, providing caretaker or receivership functions in a political vacuum, or strictly in terms of its hardware capability. These are important aspects of military rule in Latin America, but it may be more useful to hold the military factor as an independent variable. In this way the military's impact on the economy, the polity, the society, and foreign relations can be properly measured.

The Military Factor in the Economic Process

No fewer than six of the ten countries with the largest GNPs in Latin America have direct military rule: Brazil, Argentina, Peru, Chile, Ecuador, and Guatemala. Two others are essentially single-party states: Mexico and the Dominican Republic. The final two, Venezuela and Colombia, have pluralist political systems. Thus, at least 80 percent of the economic systems of Latin American countries can be said to have a strong military foundation; the military either is a direct ruler or a critical factor in the political system. More important, in nearly every country in which the military is predominant, the average annual real growth between 1971 and 1980 was significant. Even in Chile the argument can be, and has been, made that the military has dramatically improved the economic situation since Allende, having reduced inflation from an annual rate of 400 percent to approximately 20 percent in 1979.[15] At the same time, the net level of foreign debt in Chile has been dramatically reduced and real growth of capital increased, albeit at the expense of the working classes.

Military tutelaries have changed the economies of Latin America, which for so many years either atrophied or were entirely beholden to North American and Western European spheres of influence, by creating more stable conditions for indigenous capital than civilian governments had been able to achieve in the past, and thus have succeeded in forging nationalist ambitions at the economic level. The military of Brazil or Argentina has by no means adopted an open-door policy toward foreign enterprise. The tendencies are toward joint ventures, severe technology transfer regulations, and carefully worked out mineral resource and agrarian exploration contracts that can greatly benefit the nations involved. Very simply, the higher the degree of military authority, the lesser (not the greater) the extent of foreign influence and domination.

A few illustrations will indicate the highly nationalistic approach of the military in regard to who controls the domestic economy. In Argentina, very much as in the United States, the chief indicator of economic well-being is the automotive industry. After the overthrow of the nominally civilian regime of Isabel Martínez de Perón by the military junta, the extant automotive companies nearly doubled their output, from an average of 10,000 units per month under the nominally civilian regime to an average of 18,000 in September 1978, only a year later. Similarly, Argentine wheat has remained a consistently high-yield crop, as evidenced by the search for new overseas markets. At the same time, the relative tranquillity instilled by the military regime of Jorge Rafael Videla encouraged loans of over $800 million from the World Bank for electric power and highway projects, as well as private and foreign investments in oil, which in turn have stimulated a variety of national development activities.[16]

The military in Argentina, following the Brazilian pattern, has also severely tightened tax collection procedures. Only 44 percent of current spending in Argentina is financed by borrowing. Economic growth, linked as it is to the tightening of fiscal responsibility, has produced a stunning turnaround in the Argentine economy. For the first time since the overthrow of Perón, production is outstripping wage and price increases. The military government was in power for only a short period of time before these changes became evident; it is difficult not to attribute this turnaround in the Argentine economy to the military seizure of power, and its attendant reduction of the civilian bureaucracy.

The Brazilian situation is well known. Since its accession to power in 1964 the military has made a country with a negative growth rate the economic marvel of South America. The distribution of this wealth has been criticized, quite properly, as uneven and indicative of a trickle-down rather than an egalitarian conception of nationalism, but the constant, if uneven, growth under military auspices is undeniable.

Brazil has attracted overseas loans and investments of nearly $7 billion—more by far than any other country in Latin America, but carefully monitored so that control and a measure of profitability remains within Brazilian hands. In the long run Brazilian-style military rule may lead to the same kind of investment crisis faced by earlier civilian regimes. The military palliative may be good as a "shot in the arm," but the same secular tendencies may emerge under military rule as under any other regime. Brazil is now facing the sorts of problems that led in 1964 to the coup d'état: a 400 percent increase in the cost of foreign minerals (specifically oil), a serious balance of payments deficit, and increased foreign indebtedness. Internal inflation, which the military regimes kept in check for many years, is rising again, as it did in the late stages of the Goulart regime. The current inflationary rate of roughly 45 percent is higher than it was in the late 1960s. The foreign debt has risen to $24 billion, and there is some question whether, even with its current average of 9.3 percent growth, Brazil can repay that debt within the foreseeable future.[17] Indeed, the evidence indicates that debt repayment has not been possible under conditions of high consumerism and lower productivity.

The military response to economic difficulties in Brazil and elsewhere is quite different from that of civilian regimes.[18] Brazil's largest steel complex (Compañhia Siderúrgica Nacional), to which the World Bank has committed $260 million for financing, was inspected by the Bank and found to be grossly mismanaged and plagued by astronomical cost overruns. In the past the political response would have been either an effort to disguise ineptitudes or a ringing defense of the nation against foreign imperialist calumnies. The Brazilian military regime responded by simply complying with the recommendations of the World Bank. More than half of the steel company's board of directors was dismissed. The company hired an engineering subsidiary of United States Steel as a management consultant. Far from crying outside interference, the regime recognized shortcomings and absorbed criticism in a manner which perhaps only the military is capable of imposing on an economic system and its bourgeois managers.

Peru represents an interesting example of a military regime

that seized power in left-wing terms. Former president Juan Velasco Alvarado set forth an aggressive and nationalistic series of political and economic policies resting on the expulsion of foreign corporations. The Peruvian junta rejected the kinds of recommendations made by the World Bank and the International Monetary Fund which the Brazilian regime had taken seriously. But the Velasco regime came to an end, and, with it, the left-wing rhetoric. Sheer militarization of state power without building a broad-based class consensus had proved to be insufficient.[19] What then took place was the familiar coup-within-a-coup, which brought to the fore a new military leadership. Francisco Morales Bermúdez, himself a technocratic minister of finance, has sought to balance the aims of the original revolution against mounting demands for economic stability. Interestingly, during its strong left-oriented phase, Peru was one of the major South American recipients of private borrowing funds from the United States, although relations between the two nations were strained.

Since the bloodless coup, ultra-nationalistic demands have been suppressed, especially those of students. More significantly, leaders of the miners' union, who had threatened to close the copper mines, have been arrested, work output has been maintained, official state ministries have been purged, and, above all, corruption both within ministries of the nation and the private sector has been suppressed. The Peruvians speak of this as a "correction." In point of fact, this kind of controlled economic growth under military auspices is characteristic elsewhere in the hemisphere, but with much closer class cooperation than in Peru.

The Peruvian case is especially intriguing since it shows that a military capable of harnessing and orchestrating the society as a whole ultimately determines the success of a regime.[20] That this represents a renewed emphasis on trade-and-aid relations with the United States while maintaining arms purchases from the Soviet Union does not obviate the additional fact that neither the military regime of Peru nor those of the rest of South America have turned away from nationalist goals and ambitions.

We are not dealing with a mechanical, automatic correlation between military rule and economic growth. In Paraguay, the military continues to function exactly as the old civilian regimes had in the hemisphere, as a caretaker for colonial domination. Nine-tenths of Paraguay's banking capital and four-fifths of its resources are controlled by foreign firms. Beyond that, twelve of the fifteen large industrial enterprises are also foreign owned. They are in charge of 80 percent of Paraguay's foreign trade. The level of national development is so low that 88.5 percent of the firms described as industrial in Paraguay are simply artisan shops with fewer than five employees.[21] Paraguay has employed repression to prevent economic growth, not as a mechanism to stimulate it.[22] It might well be that the longer the military regimes are in office the more likely they are to behave as caretakers for export-import or *comprador* bourgeoisie. For the present economic revitalization has occurred in countries that had vigorous military coups, when the bourgeoisie, aristocracy, and proletarian classes had proved inept.

The Political Factor in the Economic Process

The military impulse is nationalist rather than socialist. Even in a country like Peru, earlier socialist and expropriating tendencies have been sharply reversed. So much has been written on this subject that perhaps we should try to emphasize why, rather than how, the military develops such strong anti-Left postures.

The political Left has taken a guerrilla posture in Latin America and elsewhere. But paramilitary units are not capable of destroying military regimes in their entirety. On a functional level, there is a profound struggle between military forces and the Left as a potential military rival. In Argentina, the military government has imprisoned innumerable citizens for so-called political and economic crimes. Imprisonment began to level off when the guerrillas seemed to give up trying to attack military bases. In Uruguay, the military regime invested a great deal of time in destroying the Tupamaros, the left-wing paramilitary group which was so powerful in the early seventies. The overthrow of the second Perón regime was more profoundly a direct assault on the Montoneros. Actual combat between the established armed force and the guerrilla armed force inevitably produces political polarization as a functional consequence of what the combatants represent in "class" terms.

The military comes to power when the economy has reached a

level of decomposition through inflation and unemployment, with resulting class polarization. This was true of Brazil in 1964, Chile in 1973, and Argentina in 1976. Civilian regimes, in an effort to survive, invite military participation in the political process. Instead of stifling or coopting the civilians, the military in Uruguay made this civil sector aware of its own lack of leadership, no less than the general weaknesses and impoverishment of parliamentarianism. The older political civilians gave the younger militarists a sense not only of participation but of what the latter could do were they fully in charge. Another crucial factor in military power is therefore the ineptitude of the civilian sector and the collapse of democratic fervor. In Uruguay, Chile, Argentina, and Brazil the military was involved in the civilian political process as a prelude to outright rule.

Because the armed forces are castelike in nature, it is difficult for them to lay the groundwork for a return to civilian control. There are few examples of a South American military regime successfully turning power back to a civilian regime. One exception was the return of power to Juan Perón by General Lanusse, but this major recognition of Argentine military ineptitude was doomed from the outset. The Peronists remained under military aegis from the beginning. Beyond that, Juan Perón was himself a military officer with a strong sense of military values.

Shared bureaucratic, organizational, and hierarchical values characterize the military, rather than any specific class connection or ideological persuasion. These shared values tend to create uniformity in military regimes throughout the hemisphere. Military dominion emerges from the infirmities of the political system from within and the ineptitude of the masses from without. Under such circumstances, the Latin American military has become both highly politicized and structurally entrenched.

The military sees itself as protecting hard-won national gains. The bourgeoisie, proletariat, and aristocracy are each viewed with suspicion, as sectional or segmented classes without the capacity to generate either an electoral consensus or charismatic leadership.[23] General Silvio Frota, in a speech marking the twelfth anniversary of the military seizure of power in Brazil, observed: "Our revolution has already done much for the moral cleansing and social and economic progress of the country. But there is still much to be done." In Argentina, General Videla urged shared effort between militarists and politicians, but not shared control of power. "There has to be a joint effort between the military and the civilians. If there is not, the civilians will simply take everything we do and turn it around 180 degrees after we return to the barracks." And President Bermúdez of Peru bluntly noted that "a person cannot be permitted to bring down the first truly revolutionary experience Peru has ever seen."[24] There is a long-term commitment to political domination and rule by the military, coupled with a deep-seated fear that civilian government is only a euphemism for class conflict followed by political chaos. If there is a new element in the equation, it is the long-term nature of the military commitment to political power and its corresponding disbelief in the older formula of managing political upheavals and then returning to the barracks.

The Sociological Factor in the Economic Process

The social consequences of military rule are harsh. With precious few exceptions, there is little point in assuming that Latin America can boast of fully evolved democratic societies. The military's goals are stated in terms of efficiency rather than democracy; they tend to encourage highly moralistic societies. Military regimes not only are repressive and harsh in terms of civil liberties and social welfare, but they also stress a vision of moral redemption no less than social evolution.

The major reason for the anti-democratic spirit is the sheer volume and persistence of guerrilla activities. While many argue that guerrilla activities may be a reaction to military rule, an alternative hypothesis is that the military is undemocratic only in its efforts to stamp out guerrilla insurgency. Whatever the causal relationship, the absence of democratic norms of legitimacy creates a climate for official and unofficial violence, leading to extra-legal societal patterns. In Argentina, internecine terror became a national way of life on both the Right and the Left:[25] mass killings by right-wing terrorists of the AAA (Argentine Anti-Communist Alliance) as reprisals for mass assassinations by left-wing guerrillas of the ERP (People's Revolutionary Army). The military regime, unless it generates developmental takeoff, can become captive to the police. In certain national contexts, the police rather than the armed forces are the true leaders of these societies. As the military regimes themselves become more diffuse, controlling terrorist activities becomes specifically a police function; and the enlargement of the scope of police activities ultimately involves either the cooptation of a military segment or, if necessary, the displacement of mild military rule by hard-line rule.

The growing differentiation, and in some nations schism, between the armed forces and the police security forces has been a major development in Latin America. In many countries, such as Brazil and Argentina, it is axiomatic that the police are more dangerous than the army and are less concerned with problems of civil liberties and individual rights. At least three factors help to account for this situation. First, strong police forces are an inheritance from an earlier age in Spanish history, when the much feared police, together with the military, spread terror throughout the rural and urban populations.[26] Second, the police function so specifically in terms of the maintenance of law and order that the least display of populist political outburst is met by repressive administrative response. Third, little or no national legitimacy is bestowed upon local or regional police forces. Police tend to be concentrated in large cities, the same places where there are major outbursts of guerrilla violence. Hence, the police function in terms of local needs and pressures, without reference to larger national or popular considerations which inhibit repressive measures.

In a major report on human rights in Paraguay, Stephansky and Alexander indicate with great subtlety this shift from military power to police power. And what they declare to be true about Paraguay can, with equal measure, be claimed for most nations of South America:

> The role of the military in the Stroessner Establishment has substantially declined, and its place has to a considerable degree been taken by the police. This has come about as a result of substantial changes in both institutions. The Army is reported to have little value as a military force only of relative use as an organ of repression. Unlike the situation in most conscript-armies, the non-commissioned officers are generally not professional soldiers, but are conscripts like the rank and file soldiers. Furthermore, the Paraguayan armed forces are poorly equipped. . . . The military also are poorly financed. . . . In contrast to the

> military, the police are virtually all-pervasive in Stroessner's Paraguay. They are engaged in the normal activities performed by any police force: directing traffic, hunting down common criminals, etc. However, from the point of view of their importance to the regime, they are particularly engaged in keeping track of the opposition and potential opposition—in harassing and persecuting its members. . . . They are very numerous, are relatively heavily armed and are professionals. If there ever were to be a showdown between the police and the military, the police might well win such a conflict.[27]

A second element involved in the harshness of Latin American military regimes is their "right-wing Stalinism." The developmental impulse is so strong among the military officer corps, and the military has become so identified with the capacity to increase economic productivity, that the sociological consequences for large numbers of people in the efforts to realize these economic goals are exceedingly harsh. For example, the trade union movement in Brazil has been suppressed; the Peronist movement in Argentina is decimated; independent unionism in Peru has been curbed; the Socialist Workers Alliance in Chile has been dismantled. Each of these results occurred in response to demands by workers for pay raises without corresponding increases in productivity. Shorter working days create huge gaps between major export obligations and the ability to fulfill them. And yet worker ownership and participation in decision making is not nearly as dangerous as the collapse of morale and the breakdown of decision making per se. Collective indecision or excessive worker demands that can be met only by the manufacture of paper currency lead to higher inflation rates and greater unemployment.

The bourgeoisie historically, and notoriously, has been unable to cope with such worker demands. As a result, the maintenance of social development has fallen to the military as the main battering ram. This means that the military imposes certain norms of output and productivity upon both management and labor. In exchange, the armed forces tend to be exceedingly repressive with respect to any violations of these norms. From the working classes they extract high-level output in exchange for highly charged nationalistic rhetoric. From the entrepreneurial classes they extract high taxes in exchange for anti-communist rhetoric and the offer of a continuation of bourgeois forms of production. From this derives the so-called moral basis of military regimes. One sees here the basis for new forms of stratification not unlike what one witnessed in the industrialization process of the Soviet Union—and with similar high dosages of violence.

The imposition of military rule has also meant greater racial stratification, even in countries such as Brazil which were formerly sociologically far less racially conscious.[28] In nation after nation the armed forces tend to be headed by whites, although the troops, whether in the Andean countries or in the Dominican Republic or Cuba, are often blacks and mestizos. Both in the southern cone of Chile, Argentina, or Paraguay, where the total black and Indian population is small, and in Venezuela, Ecuador, Peru, or Brazil, where black/white ratios are much more heavily weighted toward blacks and Indians, there is strong emphasis on white leadership and black rank-and-file participation. Interestingly, the Cuban regime exhibits a pattern of white military leadership in its officer corps very much like the other countries in Latin America, with a largely black "battering ram" to service client states in Africa. In some countries, such as Bolivia and Paraguay, the Indian population is considerable; but again, it has no representation in the military officer corps. Whether black, mestizo, or Indian, non-white groups are all outside the power vortex of the political stratification system.[29]

The same division exists throughout the rest of the social system. Although Brazil is one of the world's largest Roman Catholic nations, there is only one black archbishop. Not a single black is in a high government post; not one cabinet minister, governor, or senator is black, although there are a few blacks at the lower echelons of political leadership.[30] There is a strong tendency within the military to identify blacks with crime. The former interim president of Uruguay, Alberto Dexichiller, suggested that the United States could help stop communism by maintaining racial segregation in Africa. Argentine military leaders have expressed the belief that blacks are the cause of the United States crime problem, and that there is no street crime in Argentina because there is no black population. There is of course street crime in Argentina. But that "minor fact" notwithstanding, the military tends to prevent a heterogeneous officer corps from arising, and hence to perpetuate social divisions within Latin America based on race, while at the same time mediating the claims of classes in urban regions.

The concentration of Latin American political life within the urban sectors creates such a sharp split between urban and rural regions as to permit the military easy access to power by control of a few main capital cities. The ecology and sociology of coastal Latin America are in dramatic contrast to a profoundly underdeveloped interior, which becomes an asset in military domination. Internal colonialism becomes the essential handmaiden of military rule, maintaining sectional and class differentiation, and stimulating economic growth at the same time.

The price of the economic miracle is the kind of class and race exploitation that permits high growth, concentration of wealth, and an unequal division of profits: precisely the economic takeoff conditions that obtained in Europe in the eighteenth century and in North America in the nineteenth century. It is a historical commonplace that the price of economic growth is intensification and the maintenance of social exploitation. The military in Latin America does this with greater skill and abundance than either the bourgeoisie of Western Europe or the police of Eastern Europe were able to muster.

Developmental versus Dependency Models of Hemispheric Militarism

The greatest myth of all is that American foreign policy is controlling the military regimes of Latin American nations. It is more nearly the case that the United States has sought, mostly in vain, to prevent the worst excesses of military regimes from taking root. An examination of the current status of United States military aid to Latin America—whether in the form of direct sales by government, commercial sales by private military corporations, or credit and loan programs—reveals, without any doubt, either stability or decline in military aid programs.

The extent of the falloff in military expenditures can better be appreciated when it is understood that such leveling does not take into consideration inflation and the resulting lower purchasing power of the United States dollar. At the same time, cultural and work aid programs have been maintained at higher levels than in the past. Programs involving food and nutrition, population planning and welfare, and human resource development have all shown a marked increase over time. Some will see in this shift nothing more than a transformation of tactics from direct military colonialism. However, there are less bizarre interpretations.

The United States posture toward Latin America has steadily but inevitably shifted from support of right-wing extremist regimes to support of centrist regimes, whether they be of a military or civilian variety. The most direct illustration of this shift has been the decline of military support. Chile was cut off without any military aid whatsoever. In 1976, Congress suspended military aid to Uruguay. And beyond that, the United States allowed Peru to become the second nation in the Western Hemisphere to buy major aircraft supplies from the Soviet Union rather than encourage an escalation in arms sales to Latin America. United States foreign policy has shifted toward isolating the more rapacious sectors of the Latin American military, with some degree of sensitivity to the charges of United States military support for reactionary regimes in the past.

The question of reverse dependency is now on the agenda: To what degree is United States foreign policy captive to the military domination of Latin America? Whatever the momentary shifts in strategy and tactics of foreign aid, it is plain that Latin American militarism is defined by an autonomous set of characteristics that do not fade in the night when foreign aid ceases. Despite the termination or cutback of military foreign aid, the integrity as well as the durability of militarization in Latin America remain constant.

The unitary character of Latin American militarism and the unrelieved support for such tendencies within the socialist as well as capitalist power blocs are attested to even by those who start with fervent Marxian premises of dependency. At the conclusion of a most capable, lengthy analysis based on dependency theory, Robin Luckham is driven to acknowledge that:

> The support of socialist countries—for example that of the Soviet Union for military regimes as in Nigeria, Uganda, Egypt or Somalia—is difficult to distinguish from that of the Western capitalist countries. To be sure, it extends the political and diplomatic influence of the USSR. Yet, it reinforces the same kinds of military-political superstructures in much the same way as do

United States, British, or French military aid and arms sales to the same kinds of regimes.[31]

In a most perceptive comment, Mark Falcoff has pointed out what every serious analyst of the hemisphere knows but may have difficulty articulating because of ideological ballast: the increasing degree of autonomy exhibited by the Latin American military and their virtual imperviousness to strict ideological lineups.

> Military ambition in Latin America can assume many forms, only one of which is to curry favor with the U.S. Embassy. Some officers have striven mightily to prove that they are better patriots for having accepted Yanqui favors: Yon Sosa, leader of the Guatemalan guerrilla movement, was trained in the Canal Zone and at Fort Benning while an army lieutenant, and most of the officers of the current left-wing junta in Peru were exposed to some U.S. military education. . . . Military-leaders in other countries as well have discovered that domestic political fortunes can be vastly enhanced by Yankee-baiting; no less a reactionary than Argentine General Jorge Carcagno, who gained his combat decorations suppressing the 1969 uprising of workers and students in Cordoba, and lambasting the United States at a meeting of the Inter-American Defense Board in Washington in 1973. Even the Brazilian generals may have other tricks up their sleeves. One of them recently confessed to a West German journalist that all the foreign investment flocking into Brazil was wonderful. "We hope it keeps on coming," he declared, "because there'll be just that much more to nationalize later on."[32]

Nationalist military evaluations concerning the potential for expropriation of foreign holdings may not be correct, yet the simpleminded equation of the Latin American military with foreign business interests cannot be sustained; not in an environment in which some of these nations are now heavy arms exporters no less than importers.

Latin America is one area of the Third World where United States policy makers have moved toward fundamental distinctions between the realities of development and the rhetoric of anticommunism.[33] In countries such as Brazil and Peru, where the military makes distinctive contributions to the political, economic, and social development processes, economic aid has been substituted for military support. But in nations such as Chile and Uruguay, where military regimes operate within a doctrine of anti-communism and generosity to private investors while committing profound acts of official terror under a permanent state of siege, United States policy has been somewhat more cautious and fiscally reserved. It has been argued that the Brazilian model at least constitutes development through dependence, rather than in lieu of dependence.[34] But this sort of argument fails to take note of countervailing forces that make nations such as Brazil—and in this connection one might add Mexico—hemispheric powers in their own right, quite apart from North American wishes or ambitions.

The United States is so intent on not being viewed as the supplier and underwriter of Latin American military regimes that, although the entire world has become a military arsenal, the United States provision of hardware (generally through purchases rather than grants) to Latin America has remained constant at 4 percent of the total foreign aid program. Although United States military manufacturers have become one of the major suppliers of military hardware worldwide, Latin America does not receive these dubious artifacts: not one country among the twenty leading recipients of United States military hardware is Latin American. Total United States foreign arms sales have leveled off at approximately $6.5 billion annually, but Latin America has not received a significant portion of this amount. The capacity to purchase arms from Europe and the increased ability of many nations to manufacture their own hardware have changed the complexion of foreign aid, and with it dependency on foreign arms.

The United States has supported the military regimes, and through such agencies as the Inter-American Defense Organization it has created a climate for military exchange of information and organization. Yet the military systems of Latin America act in nearly total obliviousness to United States desires; they have their own internal dynamic. The Brazilian government is the bastion and center of military organization in South America. Under its tutelage military commanders from more than a dozen Latin American countries have joined to plan a common strategy and to coordinate their activities in stamping out what they view as a communist threat. This military group called for a second war for American independence, and it was predicated on stamping out all dissent. The United States, under past administrations, did not support this effort or its policy conclusions. It has thus far remained outside the framework of this new military group within the hemisphere.[35]

Only three nations in Latin America are indisputably capable of invading another Latin American nation or threatening another nation. They are Cuba, Peru, and possibly Brazil. It is ironic that two of these three—Cuba and Peru—are nations whose armed forces are for the most part, if not entirely, supplied with Soviet weaponry. The situation in Cuba is well known: a nation whose armed forces have become part of Soviet foreign policy. The Peruvian armed forces, based on two squadrons of Sukhol 22 fighter bombers, 200 Soviet T55 tanks, and a fleet of Sukhol 20 bombers, are reinforced by French Mirage fighters and British deHavilland bombers.[36] The American contribution to this is a fleet of fighter planes. An effect of the competition between the Soviet Union and the United States is that Third World countries like Peru develop a shopping list of hardware demands that require acquiescence under threat that they may go over to the "other side." The crucial factor is not only the degree of military hardware but the extent to which this military hardware requires a system of parts, foreign techniques, and personnel; and, in addition, credit lines that make participation in the world military system rational and feasible. But the curious fact is that, with the possible exception of Brazil, only countries supported by Soviet armaments have an international military capability.

The military leadership of Latin America is limited in what it can do. Cuba's leverage, for instance, is directly connected to the capacity of the Soviet Union to reach total economic parity with the United States in foreign aid. Although Soviet foreign aid has increased to a point where it is now over $1 billion per annum, United States foreign aid is closer to $5 billion per annum. Hence, the United States is in a better position to apply pressure to those nations of Latin America seeking new credit lines. Soviet repayment terms for goods are often stiffer than those asked by the United States. Soviet aid costs 2.5 percent interest, with repayments over a period of ten to twelve years, after a grace period of two to four years. The United States credit averages 2.8 percent interest, with repayment spread over an average of thirty years, with a seven-year grace period. The United States can provide a balancing civilian element unlike the Soviet Union, which does not have soft goods and modernization goods, and therefore has to deal in arms. Beyond that, the Soviet Union has little available hard currency reserves to offer in place of the military requirements of those countries.

It is exceedingly important that these facts be presented candidly, not to prove the superiority or virtue of the United States over the Soviet Union but rather to indicate the high degree of similarity in foreign aid priorities. Even partisans must recognize that the support of socialist countries for military regimes is difficult to distinguish from the support of capitalist countries. To be sure, such aid extends the political and developmental influence of the USSR, far beyond anything imagined by the Western powers. Yet in fact the same kinds of military and political aid and arms sales go to the same kinds of regimes. Arguments concerning the virtues of communism versus the evils of capitalism with respect to the needs of the Third World must reckon with these similarities of purpose.

Although Latin American countries may have become familiar with the political systems of socialist states, with the singular exception of Cuba they have not pursued policies of class warfare or socialist economies. Expropriation of foreign holdings has, for the most part, not been followed up with class expropriation. But when armed forces are relatively large and influential, when civilian governments have failed to deal adequately with either internal chaos or economic failure, the military has been able to perform a unique role of underwriting the middle-class economy, stabilizing the parliamentary machinery, and organizing modern society. And in the 1980s the region has shown signs of recognizing that any process of legitimation must ultimately raise the question not simply of civilian authority but of democratic reforms.

7

Globalizing War and Integrating World Systems

The twentieth century has given us the choice between indiscriminate and unrestricted annihilation on the one hand and selective and restricted war on the other. Warfare, an occupational hazard of human existence, has ever been with us. Nuclear annihilation is actually the antithesis of war: non-selective in its destruction and offering little chance of victory or defeat—in short, the negation of human history and hence of the war games men play.

The Kantian vision of a world at peace, a warless world, a federated world, is not our concern here. Indeed, it might be said that, in the past, political peace was often purchased at the expense of economic development and thus was no less a social problem than warfare. What qualitatively marks the present-day situation and gives the current period an urgent note is that conventional war is a social problem involving death, destruction, and agonies of all sorts for relatively limited segments of the human race. In a nuclear conflict that payoff matrix is qualitatively different in design, so much so that non-nuclear wars are called "conventional." Let us then note some fundamental distinctions between classical warfare and nuclear conflict:

1. In conventional war there is a victor and a vanquished,

whereas in annihilatory nuclear conflict there is either mutual destruction or such a high degree of destruction, even of the "victorious" nation, that it is not absurd to maintain that the "living will envy the dead."

2. In conventional war there is a trained cadre of men, organized and uniformed, comprising a fighting force, called an army or navy or air force. In nuclear conflict the military gives way to a scientific-technological unit. The distinction between military and civilian citizens is diminished if not dissolved.

3. Conventional war is an undertaking of some duration, which permits rallying new armies, new alliances, and changes in strategies and personnel. In nuclear conflict time is compressed; military battles are resolved in minutes; surprise, gamble, and technology replace courage, fortitude, and strength.

4. Conventional war maximizes socialization, especially in insurgency actions. Conscription, for example, creates renewed faith in collectivity. A well-defined enemy provides long-range national purposes.

Conventional war broadens popular interest in such worldly activities as military science, technology, and, recently, even the behavioral sciences. Nuclear conflict, on the other hand, maximizes privatization. A postnuclear society, living underground for a prolonged period, would snap the bonds of association by reducing people to their primitive survival impulses. The ambiguous nature of enemies in nuclear strikes confuses a populace about the true character and purposes of friend and foe alike. The impotence of the person when confronted by large-scale nuclear activities effectively destroys impulses to bravery, courage, and personal valor, replacing them with fear, retreatism, and the ritualization of aggression.[1]

Annihilation was once considered an axiomatic outcome of war, but now it is perceived as much less inevitable. The more important distinction is that between a conventional war and a nuclear war: between war as such and mutual annihilation. If the United States has 1000 silos with Minutemen missiles in place, and the Soviet Union has 1400 silos with their equivalent to Minutemen in place, either there is a Mexican standoff, albeit an economically costly one; or, if such weaponry is employed, there is annihilation, and annihilation is the end of the game. If what constitutes winning and losing short of annihilation cannot be determined, then there is a new and different game.

Nations have a concept of war, and they do go to war with one another. But they do so at sub-nuclear temperatures, in a way that does not involve the ultimate destruction of either side. An almost medieval concept of warfare as finite and limited endures: warfare conducted at some level below boiling or freezing temperatures. For this to be possible there must be a component which might be called the threshold concept, both the threshold at which the military structure is maintained and the psychological threshold beyond which one dares not go. For the war game to continue in this day and age, there must be a threshold concept that everyone observes with or without formal rules or regulations, with or without SALT II.

The question of formal treaties becomes technologically irrelevant, since issues of compliance are observable in character. At the level of military affairs there is a concept of parity, whether it is built into a strategic arms limitation agreement or not. Would SALT II result in lower military spending? W. Scott Thompson has claimed that during the first five years there would probably be increased spending; a decrease would not result until the advanced stages of the SALT agreement.[2] The United States economy's development of capital concentration at the expense of labor-intensity is characteristic of its military as well; throughout history the armed forces have been labor-intensive. The United States armed forces have ranged from four million men under arms to two million men under arms; increasingly, this country is moving to a capital-intensive concept of the military. In the short run, neither the Soviet Union nor the United States need reduce its manpower supply under SALT II; high levels of manpower can be maintained under SALT II. In the long term, there will be a downward turn of manpower after SALT II, if for no other reason than that mass opinion favors manpower reduction and voluntary participation. A guns and butter program is feasible only in a crisis. When the crisis evaporates then the guns and butter approach evaporates and yields to demands for emphasis of one or the other.

More intriguing in the war game context is the notion of hardware thresholds beyond which a nation cannot go. The current threshold indicates that neither side can use nuclear weapons of any kind; even gas warfare or napalming reaches the edge of the threshold; whereas laser systems can be deployed. Both military parity and psychological thresholds seem to be absolutely necessary to maintain present levels of tension without destabilizing or destroying the world structure.

Despite the tensions of the moment, a strong likelihood exists that SALT II or its equivalent will, with some modifications, be ratified and implemented. The level of GNP has fallen in very dramatic fashion for both the Soviet Union and the United States. Military expenditures in the Soviet Union are 20 percent of the national budget, and the social costs of such expenditures have become dangerous to Soviet survival. But the sense of parity has become so enshrined as a goal, so rooted in the relationship between the big powers, that no matter what the level of tension over a particular issue, whether it be Afghanistan or Somalia, the SALT II negotiations will be concluded. When they are held in abeyance to express annoyance or chagrin, as they were in regard to Afghanistan, SALT II itself becomes part of the war game strategy.

Industrial powers support a conventional war system that is maintained alongside an annihilation system that is disavowed and disallowed. The war system is the arrangement and the programming of conflict. It permits a variety of military behaviors that prevent any one side from becoming excessively powerful. Precisely because nations share a sense that annihilation is impossible to contemplate, war gaming remains an element of present-day conflict. It is entirely conceivable that there could be a major nuclear war; the hardware is available, and at certain levels sufficient provocation could occur. A high degree of communication between powers maintains a low probability of violence. Like the Middle Ages, the postindustrial era is an arranged environment—an arranged hostile environment.

The question of war does involve an ultimate survival question. Annihilation questions are ultimate. That is why they possess a sordid existential property. If the major powers are intent on destroying the other side, what would the world look like? By the same token, if two people play Monopoly and one party starts winning heavily, the other might take out a gun and shoot his opponent dead. Although such a "scenario" could happen, the presupposition has to be that while one party may win this game of Monopoly, the second party could catch him and win a later game. To assume ultimate recourse to ultimate weapons is to talk about the prospects for annihilation, not war.

Today there is a stable condition of neither peace nor annihilation. But there is the possibility of annihilation by accident. That is why mutual force reduction and bilateral rather than unilateral concepts of war and peace continue. The military is in charge of a war system. Now if one side absolutely denies the possibility of losing and the other side has a complete monopoly of power, the world as we know it will come to an end. At the moment, unilateral force reduction, or unilateral termination of military expenditures, would result in one side, the Soviet side, prevailing.

The SALT II treaty permits some specific disequilibrium remaining, but ensures overall parity, with geographical-demographical distribution along hemispheric lines. Will SALT II be attractive or ethical? Will small powers be happy with it? Probably not. SALT agreements are not intended to create a climate of peace. They are intended to create a climate in which fear of annihilation can be overcome. Nothing in the SALT agreements will eliminate problems in Zimbabwe, Somalia, the Caribbean, or Southeast Asia. Low and intermediate threshold issues are basically not addressed by SALT. In this sense, such treaties should be viewed as ground rules that both limit and permit warfare.

The substance of the matter is that war games persist because relative advantages accrue to those who play them vigorously. The unique feature of the 1980s, in contrast with the 1960s, is that the strength of the players has profoundly altered over time. One can now say that the Soviet Union has a firm lead in a wide variety of nuclear as well as conventional armaments. But how does the United States "play the game" under such drastically altered hardware conditions? The burden of this chapter is to demonstrate how the war game is programmed in an epoch of a changing balance of military forces, which in the short run, at least, favors the Soviet Union over the United States.

Because wars are arranged, the possibility of intense struggles is omitted or alleviated. Direct confrontation between the United

States and the Soviet Union becomes an extraordinarily serious event. The Third World, on the contrary, provides low-intensity areas, such as Vietnam or Kampuchea, where proxy struggles take place, e.g. between the USSR and China. Another example of a proxy struggle is the Lebanese Left versus the Lebanese Right, when the real struggle is between the PLO and the Israelis. In the Angolan civil war between the MPLA and UNITA, there is a confrontation basically between the USSR and the United States. In Ethiopia versus Somalia, Ethiopia is the Soviet proxy and Somalia is increasingly becoming the American proxy. All proxies are risky and spill over in one direction or another—Somalia was formerly a charge, if you will, of the USSR. But the very fact of a kind of switching at low-intensity areas is itself indicative of the feudal character of contemporary warfare. Low intensity requires a high level of communication. The higher the level of communication, the lower the intensity of the struggle. This kind of situation is exemplified in terms of hardware in the system. The big hardware expenditures systems are the USA and USSR. Between them, 85 percent of the world's hardware is dispersed. The smaller players of the First World, such as France or England, also control substantial arms. For example, the United States is currently arming Egypt with F-16 jet fighters for the first time, an advanced weapon system that will give it parity in relation to Israel. It will also provide for interchangeability of parts in that area, vis-à-vis other participants in the Middle East cockpit. Kuwait, formerly an ally of the United States, is now being armed by the Soviet Union. In terms of the recipient rather than the donor there is a play within a play. Peru, for example, will buy airplanes from the Soviet Union, tanks from the United States, and ammunition from France. In other words, it is able to manipulate the present parity arrangement between the First and Second Worlds to its own tactical advantages. Certain nations are adroit at this game by virtue of historical circumstances and natural resources.

The war game at some level involves the empirical determination of and a gamble on world economic systems. There are a series of metaphysical assumptions disguised as historical inevitabilities or moral superiorities that the best system will prevail. There is an ideological game that goes on at abstract levels as well: the Soviet people and leadership believe in the superiority of their system, and the American people and government believe in the superiority of their system. The measurement of alternative systems in terms of how much is produced and for whose benefit ultimately will determine the outcome of the game. The problem is that levels of productivity do not correlate with forms of distribution. Games are not necessarily played by those who are in perfect equilibrium. Games are not played between those who have and those who do not have, between those who are more innovative and those who are less innovative. Games can be viewed as a contest over the superiority of systems—our system versus their system, our athletes versus their athletes. In a condition of intense competition the game metaphor becomes very real. Now one can suspend the real world game and say "I don't want to play the game anymore, let's blow it all out." One always has the option to terminate by destruction, but then the game is over, the world is over, and what has been discussed ceases to have any meaning.

The Soviets are probably as frustrated as the Americans by this overriding, paramount fact. In some respects, the game metaphor concerns programming rounds and programming stations of conflict. It doesn't necessarily mean that there are no winners; it means that the game is long term. And over time games employ longer and longer plays, allow more and more players; they increasingly take whole days to win. New equipment is required. The new games are very complex; they involve rotation of positions and players. Going back to a simplified game has consequences that are extraordinarily risky. One has to believe that at some level the rationality of self-interest emerges; what keeps the game going is a feeling of mutual superiority. The Soviet Union is as sincere as we are in touting the merits of its system. If they are right they will win. If we are indeed flabby and weak, we will lose. Ultimately, there are winners and losers.

An exceptionally significant change over the past several decades has been what might be called a growing methodological similarity between the United States and the Soviet Union. At an earlier stage, the Soviet Union was unable to break the shackles of its mechanical Marxist formulas, and this was a key danger. Their earlier formulations had an air of myopic bravura: the United States because of its enfeebled socioeconomic system would perish in an atomic war, but the Soviet Union by virtue of the socialist unity of its people and purpose would survive such a catastrophe. The imbalance between pragmatic American and dogmatic Russian formulations, disguised with ideological overlays of supremacy, represented a special danger, since gaming analogies are useful only to the extent that both sides (or contestants) have a clear appreciation of the rules, limits, roles, and outcomes of any particular game of war.[3]

This asymmetric condition has radically altered with the passage of time and the erosion of ideology in the East. A distinctly friendly voice put the matter of Soviet policy aspirations thusly:

> The Soviet leadership has become fascinated with the game of geopolitics. The Soviet press gave great attention to the American military buildup in the Middle East during the hostage crisis. Before the Afghanistan invasion, scholars drew an analogy with Mexico. What, they said, if someone had seized the Soviet Embassy in Mexico City and demanded, say the return of some Mexican Communist? What if, in response, the Soviet had assembled an armada off the Mexican coast and threatened a blockade or even a punitive military action? Why, they asked, do you think our [Soviet] leaders will react differently to a buildup of military force off a country on our southern border?[4]

Less important than the specific reasoning behind such an approach is the approach itself. That is to say, the explanation that the invasion of Afghanistan was an effort to rationalize Soviet Middle East interests in the name of "counterbalance" marks a huge step beyond an earlier decade in which the imagery of dialectical "railroads" was used to justify Second World expansionism.

If in fact the game metaphor is the universal *lingua franca* of geopolitics, a metaphor that the Soviet Union has incorporated into its own strategic thinking, this very fact offers new prospects for nuclear peace and is a considerable step beyond the two-ships-in the-night situation which obtained in the Stalin-Khrushchev period of 1945–65. In that period American Pentagon thinking increasingly came under the influence of game theory models, while Soviet Kremlin thinking persisted in a nineteenth-century historicism that postulated an inevitable victory of socialism that was divorced from empirical realities. But this does not mean that a war game model eases non-nuclear areas of international struggle. The fact that war games are conducted at similar ideological and intellectual levels does not necessarily signify similar condition at the functional or military level. The last two decades have witnessed a huge growth in Soviet military capability. This growth has basically been noted but unanswered by American military evolution. This is a subject widely discussed and examined in the professional literature, and needs no further amplification. What does need to be emphasized is the global implications of this heightened war game atmosphere.

War games are primarily military games; in this sense, the United States, while far outstripping the Soviet Union in economic output and political openness, is nonetheless faced with a formidable adversary at the military, or gaming level. Richard Pipes appropriately reminds us that the Soviet Union has always devoted a disproportionate share of its resources to the maintenance of its armed forces:

> Of all the instrumentalities at the Soviet Union's disposal, it is the military that occupies pride of place. Soviet imperialism (that also held true of Czarist imperialism) is a military phenomenon par excellence, and in proportion as Soviet combat power grows, both absolutely and in relation to the West's, it tends to push into the background the political manipulation on which the regime has had heavily to rely earlier. Increasingly, Soviet spokesmen call attention to the shift in the military balance in Russia's favor as a decisive fact of the contemporary world.[5]

A critical factor in the current stage of the war game is the real and open-ended advantage the Soviet Union now possesses in the area of armed forces which are trained, deployed, and combat ready. It might well be that the Soviet advantage is transitional and will not live out the decade. But the special situation in which Soviet power is first and United States power is second is one that can scarcely be overlooked in the planning and policy headquarters of either major power. In this sense, the maturation of Soviet military might between 1960 and 1980 has encouraged its participation in war game scenarios. Far from being a monopoly of the Pentagon, sophisticated military hardware is more realistically a monopoly of the Kremlin.

The asymmetry of real military power has created conditions for symmetry in the strategic and tactical realms. Hence, the coexistence of war gaming, far from providing a sense of geopolitical relaxation as a function of strategic balance, has the reverse effect: it makes the Soviet Union a more vigorous and active player in the game of war precisely because it realizes that advantages—ours in the past and theirs in the present—may well be transient in character. Edward N. Luttwak has addressed this matter in frank and meaningful terms:

> The lineaments of great danger are easily recognizable in this situation: if an edge in military power can normally be expected to encourage activism (but also make it less necessary since others will pay power its due in their diplomatic concessions), a transitory advantage is apt to make action positively urgent. Since the elements of military power are mostly rather perishable, equipment becoming obsolete and training dulled in a matter of years, the Soviet Union needs a great effort merely to maintain the high quality of its present forces, which have been made much more ambitious in quality as well as somewhat larger in size over the last fifteen years or so.[6]

The crystallization of the Third World led first of all to the strong opposition of China to Soviet "hegemony" within the Socialist bloc; within the Middle East a movement from modernization to fundamentalism also produced concern and fear on the part of the Soviet Union. The Soviet invasion of Afghanistan is a function of an inability to cope with not only the rise of the Third World and the rise of the Middle East as a part of the capitalist bloc but the use of traditionalism in the maintenance of a Third World posture. Islam provides a highly traditional worldview that does not make very clear distinctions between state and religion. On the contrary, the foundation of Islamic thinking is the integration of political and religious structures. Unlike Christianity, it has no theory of rendering unto Caesar what is Caesar's or rendering unto Christ what is Christ's. The Soviet Union faces much more serious difficulties with respect to Afghanistan and Iran than with Eastern Europe, where classical distinctions between the religious and the secular domains obtain. The fear not only of nationalism in Afghanistan but of spreading fundamentalism and anti-modernization served to force the Soviet Union to define by military action the limits of the Muslim world. There are dangers and risks to an invasion of Pakistan or Iran or Yugoslavia, but what has too easily been overlooked is the impact of this direct confrontation of the Second World with the Third World.

The rise of fundamentalism, of a kind of religious-centered state authority, has threatened the Soviet Union more ideologically than militarily. The Afghanistan invasion raises a number of issues. The invasion could not have occurred without the express approval of the Kremlin leadership. Since Soviet Union leadership is primarily bureaucratic and technologically oriented, its level of ideological fervor is minimal. How important are Marxist premises to Soviet decision making? In Afghanistan, Third World fundamentalism was moving toward anti-Marxism. One element in Soviet thinking was the possibility that this might extend to the Soviet Muslim population. Since the Soviets operate under a conviction that once a country is in the orbit of the USSR it must remain in that orbit, the need for action against a Third World country became inevitable. The last country to escape from that orbit was Yugoslavia, and then only partially, and in historical circumstances of a post-World War II epoch considerably less favorable to Soviet power than the present epoch.

The invasion of Afghanistan by Soviet armed forces represents not so much a stepping stone as an expression of frustration with a Third World that has consistently refused to become part of the socialist march of history. Increasingly, the Third World is becoming part of the capitalist environment, especially of the world monetary system and the international capital market system. This is admittedly a special reading of the present struggle between the First and Second Worlds, but there is a great deal to recommend its authenticity. The problems that each bloc has with its own allies make direct USA-USSR military confrontation just about impossible. Bloc solidarity hardly exists. Rumania is as much a wild card within COMECON as Italy is a wild card within NATO. What has happened is a sharper demarcation of lines of influence and a withdrawal of geographical and demographical lines containing or limiting spheres of influence in nineteenth-century terms. There have been areas of penetration of each major power's sphere of influence; and they have been strongly induced by the consequence of any precipitous movement of the Third World to one or the other spheres of influence.

An enormous number of changes have taken place within the Soviet Union. An essentially totalitarian regime has been transformed into a less centralized, basically authoritarian regime. A highly stratified consumer society has emerged in the Soviet Union which was not in evidence at the end of World War II. It is a country that has not had a war for many years, as witnessed by how poorly they gauged the consequences of Afghanistan. American foreign policy, for its part, grew more aware of its limits in the 1970s, of where the lines are drawn and the inner consequences of drawing the lines differently. To some extent, the Soviet Union is interlocked with the Western economy. It has powerful banks in Zurich; it wheels and deals within the international commodity market along with the West; it has developed a businesslike style of work. The Soviet Union has bankers, businessmen, and a managerial class. The United States is not naïve, and the Soviet Union is not particularly brutal or vicious. The game is complicated, played out in troughs and peaks, lows and highs.[7] If it were a uniform game, it would not be a real game. The prospects for war in the 1980s are very low; instead there is a kind of informal Congress of Vienna in which SALT II will spill over to SALT III: it is equivalent to the political demarcations of lines of authority. The stability of this system is difficult to calculate. The identity of the players involved is important. Since 1945, we have been talking about war essentially between the two players who have real international power. In the absence of wars between these powers, if a third nation, China or South Africa, becomes very powerful, then the character of the game will obviously change. The introduction of more than two players makes the game more problematic, and beyond that the change in the players themselves has an impact on the change in the rules by which players play.

The proliferation of nuclear weapons among nations which have not previously had them is probably an overrated problem. Having an atomic bomb and having the capacity to deliver it on target are qualitatively different. In terms of a 1980 series of weapons—laser systems that can destroy a target within a matter of seconds, or bombs that come out of random fields—the only two nations that have those capabilities now or in the foreseeable future remain the United States and the Soviet Union. Pakistan might maliciously and wantonly use a bomb against India, but it is doubtful that the result would be a new world war. Pakistan probably could not deliver that same bomb to the Soviet Union. Nations may have nuclear capability but not the ability to generate the systems needed to target such weapons. What must be examined carefully are differences in delivery systems which are really First and Second World monopolies. The capacity to manufacture low-yield nuclear weapons is now possessed by eight or nine countries, enough to spark limited warfare but not a total holocaust.

The United States and the Soviet Union both have a very sophisticated level of weaponry. The laser age will initiate yet new changes. Laser beams will find nuclear missiles and explode them in air. At that point, whoever has the technological advantage in the 1980s will have enormous military advantage. That party will be able to deliver first to the target and prevent the other side from delivering its weapons on target. In the current round of the war game we are dealing less with questions of weapon quantities than with weapon systems. The advanced technology of annihilation has moved from a crude "labor-intensive" phase partially dependent on human will to a "capital-intensive" phase largely dependent on the technical capacities of each major contender.[8]

If the optimal results of a world war are a fundamental redistribution of the sources of energy, food, and wealth, then World War III has been fought and the Arab Middle East portion of the Third World has won. A redistribution of wealth has taken place as energy prices have escalated dramatically inside of ten years. This "World War" was a rare event, a world redivision of power without overt conflict, based on the monopolization of a fundamental source of energy, despite backwardness in technology or military capabilities. Historically many nations have dreamed of such a monopolization of a much-needed resource. Brazil thought that through rubber it would monopolize a source of power, but the rubber plant could be reproduced in abundance in the Dutch East Indies, and after that polymers came along so that was the end of that monopoly. Sugar played a similar role for Cuba; but there are many substitutes for sugar in foods of one kind or another. Even precious metals of one kind or another—zinc, lead, and so on—are substitutable.

The energy situation is unique. Technology fell so badly behind raw material supplies that the First World was unable to cope with monopolization of energy sources. That rare event allowed for redistribution of world wealth without armed struggle. It is hard to envision this same sort of shortfall happening again in the near future. An important factor has been the relationship between a private monopoly maintained by the so-called Seven Sisters and a government-led monopoly such as OPEC. The multinational posture created a monopoly on a world scale that no single nation could break, not even the United States. OPEC took over the decision-making power that had been consolidated by the Seven Sisters and exercised it as a mixed strategy based on a pseudo-partnership. Forms of monopolies based on energy sources are limited in terms of their potential; this one has had a longer staying power than any previous monopoly because the profit picture has been changed but not destroyed.

Could the United States take military action to break the energy monopoly? In 1973, when the oil embargo took place, the United States probably could have taken over a country like Kuwait without risking Soviet intervention. Such a strategy might have avoided the escalation of prices and costs of oil and energy, and would have bought time, without a doubt. But military action would not have addressed a serious structural imbalance. The U.S., with 2 or 3 percent of the world population, consumes 30 to 35 percent of the world's resources. At some point, a redistribution of world resources would have been unavoidable.

Despite the quick pace of change, the same players prevail. In 1903–4, who were the players? The United States, Russia, Japan, Continental Europe, and the United Kingdom. In the twenty-first century there will be other Third World players: Australia, Brazil, China, perhaps South Africa, Nigeria. With five such new players in the game, a change in the character of global power is inevitable. But what is stunning about the twentieth century is that global politics have been dominated by the same nations despite the incredible velocity of economic change. Despite the rotation of power, the wars that have been fought, the blood that has been shed, those who lead at the end of the twentieth century are remarkably similar to those who led at the beginning of the twentieth century.

The relative weights are different between the Soviet Union

and the United States, or Western Europe and the United Kingdom. In this period of rapid change, including the emergence of a Third World, the pre-eminence of the same players is a matter of no small fascination. By 1985, there should be greater parity between the First and Second Worlds than obtained in 1980. The biggest problem right now is that the United States is the second ranking military power, not the first, and will probably remain so for the next decade. The United States is incapable of delivering the munitions or the men to fight a two-front war. For the next few years the central struggle for the United States will be recapturing military parity. It is doubtful that the new players of the Third World will affect the bipolarity of the game as such; but their leverage in such a delicate balance of terror is greater than it was in the early part of the century.

A military strategy might have bought time to develop an alternative (non-petroleum) technology. The counter to such an argument would be that the time would not have been used to create a new technology; it would have been used to build larger automobiles and would simply provide a longer lead time for added consumer waste. It is an abstractly reasonable premise to say that we could have bought seven years; we could have minimized prices by having a military conflict. However, a hard-line strategy works for a few years at the most. Normal action must ensue, and the pressure on the environmental system would remain intact. The war game could have been played in terms of threat mechanisms, but a better way was in terms of technological innovation. Richard Nixon's approach in the winter of 1973–74, the first embargo, was extremely conservative. He felt that by 1990 the United States would require self-sufficiency in oil, and that all efforts would be bent in that direction. The gamble was taken to embark upon a technological rather than military route: to exercise the technological option rather than the military option. But the United States has not achieved a technological breakthrough. On the contrary, the consequence of avoiding armed conflict at all costs was that the OPEC nations were firmly integrated into an expanding world economy, even if they lacked a corresponding military power.

The Soviet Union plays the war game for the long pull of time: East Berlin in 1953, Hungary in 1956, Czechoslovakia in 1968, Poland (without invasion) in 1980. It is not a matter of changing

Soviet leadership. Soviet foreign policy has very little to do with Soviet domestic policy; for example, during the years of Stalin's internally repressive regime, foreign policy in the Soviet Union was highly restrained. There was some question about how the Soviet Union would behave after the death of Tito, but the answer so far has been prudential. Milovan Djilas suspects that it will not be possible for the Soviet Union to invade Yugoslavia. What might occur is a play on the national eccentricities of Yugoslavia. Croatians, Macedonians, Carpathians, and Slovenes are in an uncomfortable union held together by the armed forces. Yugoslavia is a country with six different national types; whether it will hold together is hard to say. It should be recalled that Yugoslavia is the only nation to develop an endemic socialist regime as a result of its own national liberation efforts. Every other country in Eastern Europe is essentially a protégé, having been invaded by the Soviet Union (or rather the Soviet Union swept eastwards toward Germany and these nations were in its path, many of them crippled fascist regimes like Rumania and Hungary). What will happen now that Tito is gone is anyone's guess. There are no successors, certainly not to his role as a titular statesman of the "non-aligned" bloc. Countries like Yugoslavia are very much like Franco's Spain; they have been involved in illegitimate rule for so many years that biology rather than politics ultimately determines when change occurs. This was the case in Portugal, Spain, and Greece, and it is now the case in Yugoslavia. The country cannot remain stable, not necessarily because of the national question of the six different ethnic rival groups but rather because its legitimacy has never been fully established. There are countries that have no institutionalized pattern of succession. Like an organization that has only one person in authority, the rules of political succession themselves become questionable. A "socialist" country like Yugoslavia has the same problem that a "fascist" country like Franco's Spain had; and that Cuba will have ten years from now. When the originating figures die, are vanquished, or assassinated, crises result. To predict the consequences of crises, or even how they are made manifest, is not possible.

Now that the economic phase of World War III is over, fought mainly in terms of how much wealth has been transferred from the OECD countries to the OPEC countries, one must note its peculiar consequences. In the post-World War III climate, these new leaders are being integrated into an advanced capitalist economy. The problem becomes distribution: What do you do with all that new wealth? On the one hand, the population in the Arab Middle East is largely underdeveloped and consumer needs are minimal; on the other hand, a very large amount of the capital has been accumulated in a very short amount of time. Hence, the process of wealth distribution has not kept pace with capital accumulation. The Iranian debacle indicates the danger in this imbalance.

Even after the oil-producing countries in the Middle East bluffed the West and won World War III without armed intervention and armed struggle—they monopolized all the energy resources—they had nowhere to go. The Arab nations had to join the Western camp although the West lost the war. There have been consequences. Arab leadership is sophisticated. It knows all about ownership, about proprietary rights. Arab economic ministers have created a pool of wealth that requires heavy investment; and the risks are great—more from internal dissension than from overseas adventures.

The gamble taken by the West in 1973 was to adopt a technological rather than a military strategy; as a result the number of players within the world capitalist arena expanded, along with heavy inflationary pressures. The oddest economic outcome of World War III is the strengthening of world capital formations. It has been at best a Pyrrhic victory for America. The redistribution of world resources means that American capitalism is weaker than it was even five years ago, and certainly far weaker than it was ten years ago. Every major indicator also shows that the redistribution process may weaken American capital but strengthen Third World capitalism. The Middle East has become a high purchaser of commodities and all the consumer products associated with an expanding economy. The redistribution resulting from this symbolic war, involving crucial segments of the Third World, has strengthened not the socialist system but rather world capitalism.

Because of this major development, one can better understand the frustration that has been experienced within the Soviet bloc. According to all canons of Marxism-Leninism, the Third World would ultimately yield to socialist outcomes. National liberation was but a first stage in the post-colonial environment. But such an evolutionary pattern did not take place. Rather, most Third World nations have been integrated into the open market system of the capitalist West. At the same time, while the Third World exhibits some rather startling growth patterns, albeit of a selective and uneven sort, the economies of most Eastern European COMECON nations are at no-growth or slow-growth phases. The Rumanian, Bulgarian, and Polish economies are each experiencing disarray and stagnation. The Soviet Union has markedly deteriorated in its ability to deliver economically or socially. Its rate of growth to achieve economic parity with the First World by the end of the century would have to be 12 to 14 percent per annum over the next two decades. A more likely level of growth in the Second World would be 2 to 3 percent per year. The Soviet system is unable to deliver on its maximum promises; it cannot sustain high levels of consumer productivity or match the industrial productivity of the West. Even if it could, in specific areas, such as electronics, the United States has a marked advantage.

In the area of miniaturization, computers, management systems, and professional services, the United States far outstrips the Soviet Union. In this particular realm the advantages are sufficiently great and the movement toward capital-intensive military hardware so definite that the possibility of the Soviet Union taking over and leaving the United States behind seems remote. Other than its direct military superiority, the Soviet Union is not in a particularly favorable situation vis-à-vis the United States. The possibility of the Third World moving into the Soviet orbit is minimal; changes occur—democratization occurs—but the process of democratization, especially in Latin America or Africa or the Middle East, will occur as economic disintegration occurs within the Eastern European bloc. Poland, Yugoslavia, and Rumania are interesting illustrations of such tendencies toward pluralism even within the Second World. In the next few years there will be a growth of Western strength because it will have capital-intensive growth in the Third World to draw upon, for example, in areas where the West is clearly superior, and willing to exercise that superiority.

The rate of growth of the Soviet economy and the rate of growth in Eastern Europe is considerably slower than that in the

West. Their starting points are much lower than those in the United States or elsewhere in the West. The disarray in the Eastern European community—the difficulty that the Soviet Union has had in stimulating innovation, satisfying consumer demands, offering adequate health and welfare care, creating an environment that is inhabitable—represents an insurmountable obstacle in short-run terms. However, problems of the Soviet economy internal to its own structure are even greater. The situation is not as bleak for the United States as it seems on a first view. Wars can be won with military troops, but after the war is over one must establish a viable system, if for no other reason than to retain hegemony in one's own region. This search for viability in economics, legitimacy in politics, and equity in society has become the preoccupation of East and West in tension, and North and South in dialogue.

A kind of Western conceit has led us to believe that the problem of development is uniquely the preserve of the Third World, whereas the problem of war and peace equally defines the province of the First and Second Worlds. Nothing could be further from the truth. The question of war as practice is precisely the shared legacy of the Third World—not as something for the postponable future, but as an omnipresence. Each of the eight violent conflicts in the world during the 1970s were fought in Third World nations. By far, the most far reaching has been the Kampuchean civil war, in which estimates of death range from between 500,000 to 4,000,000 (or no less than half the Kampuchean population). The Afghanistan civil war, begun only in 1978, claimed the lives of between 100,000 to 250,000 people by 1980. A similar number of deaths resulted in the East Timor war. The bitter wars in Lebanon, Vietnam, the Philippines, Guatemala, and Zimbabwe each claimed the lives of anywhere from 20,000 to 50,000 lives.[9] The sobering conclusion is that war may be a big power "game," but it is one played out by small-nation "players."

8

State Power and Military Nationalism

If there are "laws" of development for the Third World as a whole, there are limitations to such laws governing patterns of change for Latin America. The very age of Latin American nations, in contrast to the ages of other parts of the Third World, would make such exceptions worthy of careful examination. A seemingly endless list of naïve descriptions, masked as explanation, has been adduced. Historians, the first of the analysts, posited the sloth of the indigenous Indian population; the ruthlessness of Conquistador invaders; subsequent racial miscegenation; forced absorption of the native population into the Roman Catholic church; the immediate gratification demanded by Catholics in contrast to Northern Hemisphere Protestants said to opt for postponed gratification. Then came a barrage of sociological theories: the fact that the continent was later in its economic development than Europe; the tropical climate which kept Latin America from the mainstream of economic development; the fact that the area was settled along its coastal regions and its interior areas left in a backward condition; because primarily linked to Spain and Portugal, Latin America remained outside the grip of the most advanced industrial nations of Europe and also outside the orbit of formal democratic styles of rule. All these theoretical explanations were intended to demonstrate that

Latin America is exempt from general laws of social and economic development.

Some of the variations on the theme of exceptionalism are too idiosyncratic and others are too insignificant to repeat. All lead to the conclusion that Latin America does not follow the general laws of development characteristic of the Third World. Area specialists have supported arguments for Latin American exceptionalism in some measure because, like all who work in a foreign setting, they want to feel that there is something unique about that area, something quite different from the rest of the world. Furthermore, Latin American studies in North America evolved quite apart from and antecedent to Third World studies.

In recent years researchers have tried to reintegrate Latin America with the rest of the world. One such scheme, the dependency model, is strongly influenced by a neo-Marxian economic perspective. This approach, in its broad outlines, holds that in essence, Latin American underdevelopment has little to do with any of the aforementioned internal factors, and has even less to do with exceptional geographic aspects within the continent. Dependency theory holds that Latin America is best explained in terms of the economic needs of imperial powers.[1] Latin America is an underdeveloped area, a case study in backwardness, because of the reckless requirements of first European and then North American colonialism. Underdevelopment is *caused* by external imperialist relations. The mineral and monetary needs of North America and its trade partners in Western Europe define Latin American backwardness. The dependency model claims that Latin American stagnation must be understood in the context of North American and Western European capitalism. More sophisticated versions of dependency theory admit the possibility of real economic growth within this system, albeit retaining the idea that the cosmopolitan center in the long run limits as well as determines the autonomy of economic growth in Latin America.[2]

The dependency model represents an effort to break out of the double bind of Latin American exceptionalism. To that degree it has proved to be an extremely important and worthwhile enterprise. The model incorporates such a global approach that it provides a great deal of information about North American and Western European capitalism, but ironically, very little about what is going on in Latin America in any particular time or place. If dependency theorists consider internal class and political relations in Latin American countries at all, they see them as structured primarily by the forces of external capital.[3] Dependency theorists present a sort of metaphysical perspective which ends up as a variation on the theme of exceptionalism. In this case, the source of such exceptions is located on foreign soil. Unfortunately, the approach deflects theoretical generalization from concrete research.

An ironic consequence of the model's search for universality is that it makes precious few provisions for national differences. The pressures and interests of external capital alone cannot explain the different economic and political histories of Mexico, Brazil, Chile, or Cuba. Unless dependency theorists can explain both the Allende phenomenon in Chile as a means of foreign capital control and the counter-revolution against Allende by the same dependency theory, such a level of macro-theorizing does not come to terms with the issues at stake or offer much promise for predictions of future events. Growing awareness of this fatal flaw in the dependency model has already reduced this new orthodoxy to fragments, its intellectual stock dropping quite dramatically.

There is a practical need to return to the study of Latin America as *sui generis*: what the hemisphere represents, not simply through the prism of North American and Western European imperialism but in itself and in relation to North American and Western European nationalism and the rest of the Third World. The dependency model has proved to be just another addition to the wastebasket of theory rather than the key to a storehouse of knowledge. Instead of providing a new way to deal with Latin America and a new understanding of its relationship to the Third World, dependency theory is only another twist of the imperial lion's tail. The foreign dominion theory has an emotionally gratifying bluntness, but such global visions do not yield useful knowledge of the huge area known as Latin America.

We are compelled to return to fundamentals. The doctrine of exceptionalism does not yield a general theory to explain events. The more exceptional cases there are, the fewer opportunities for establishing a rule. There presently exists an atomistic situation in Latin American studies: twenty political theories for twenty nations. If a different theory is needed for every nation, the absence of parsimony will lead to the substitution of nationalism for science as a basis for political analysis.

Emphasis upon nationalist explanations is a major defect of current studies of the area. Social scientists rarely discuss Latin America; they prefer to deal with the exceptional situation of Chile, or the uniqueness of Brazil. This is not to deny that every nation has individual aspects. However, to accumulate exceptions is no way to construct theory. The national factor does indeed organize the economic, legal, and linguistic aspects of social life. The question for Latin America is: Does any higher principle organize the national aspects of social life?

Individual cases of development and underdevelopment can be explained only by considering both a country's relationship to the world economy and internal variables, particularly the relationship of the state apparatus to the class structure of that society. Economic development in all Latin American countries is structured by their position as late-developing states. They are trying to industrialize in a world economy shaped and controlled by those countries—capitalist and socialist—already developed. The opportunities for external exploitation and expansion, which were vital ingredients for economic development in the United States, Western Europe, Japan, and the Soviet Union, are limited and in many instances non-existent in Latin America. Moreover, in less developed states the industrial bourgeoisie is weak, and capitalization of land holdings often fails to occur. As a result, the state must break down class barriers to industrialization and either create, greatly strengthen, or substitute for an independent entrepreneurial class. Because of restrictions upon external expansion, the state must foster internal mobilization of capital and human resources. Thus, the later the stage of development—whether capitalist or socialist—the more necessary is state direction and control of the economy. The state was more important in German economic development than was the case in Great Britain or the United States, and it was even more important in Japan, Russia, and China.

The state apparatus that has the potential for playing an innovative economic role is bureaucratized, centralized, efficient, and, most important, autonomous. A bureaucratic state apparatus can be considered autonomous when those who hold the highest civil and military bureaucratic posts satisfy two conditions. First, they are *not* recruited from the dominant landed, commercial, or industrial classes, nor do they have personal vested interests in the dominant means of production. Second, such bureaucrats are not controlled by or subordinate to a parliamentary and/or party apparatus that represents the dominant interests. Parliaments and parties may exist as long as they do not control the military-bureaucratic elite. Indeed, military-bureaucratic elites often rule through parliamentarians so long as the latter do not impinge upon their political autonomy and power. Such an autonomous state apparatus is nearly always authoritarian; but not all authoritarian states are autonomous. An autonomous state, however, does make possible a clear distinction between state power and class. This conception of state autonomy is distinct from those proposed recently by a Marxist structuralist and by a Parsonian functionalist. Both approaches define and discuss an autonomous state or policy; but in a way that is contradictory and inadequate. Let us consider their analyses in order to clarify the definition presented here.

In *Political Power and Social Classes,* Nicos Poulantzas defines two ways in which the state is autonomous in relation to class forces. First, and most important, he claims that under the capitalist mode of production it is functionally necessary for the state to be structurally independent of class forces in order to maintain and protect the interests of the capitalist class.[4] Poulantzas postulates that the capitalist state always functions independently, no matter who staffs it, except under very special circumstances. For Poulantzas, another, but more unstable, state autonomy occurs where there is a balance or equilibrium between competing classes or competing factions within the dominant class so that none is dominant.[5] Only under these limited circumstances does the state cease to function as a political organizer of the capitalist class. Thus, the state apparatus rarely has power independent of social class power.[6] As Ralph Miliband says in a review of Poulantzas's book: "Poulantzas' failure to make the necessary distinction between class power and state power . . . deprives the state of any kind of autonomy at all and turns it precisely into the merest instrument of a determinant class."[7]

In contrast, my framework assumes that control of the governing apparatus is an independent source of power just as control over the means of production is a source of power for a class. Whether those who control state power are independent of or closely tied to those who exercise control over the means of production can make a great difference in state policy. It is important that those who control state power are personally committed (by vested interests) to the present organization of the economy.

In *Political Order in Changing Societies*, Samuel Huntington defines autonomy as one of four characteristics of a modern political system. An autonomous political system, he says, is "insulated from the impact of non-political groups and procedures."[8] A political organization that is the instrument of a particular social group—family, clan, or class—lacks autonomy. Yet Huntington views political parties as the most important instruments for the creation of a modern and autonomous polity. For Huntington, political parties are most successful in aggregating common interests and overcoming narrow self-interest.[9] Huntington assumes that all political parties are autonomous; he never considers that many modern parties articulate the interests of the capitalist class in such a way that they seem to express a general interest without in fact doing so. Miliband says: "The conservative parties, for all their acceptance of piecemeal reform and their rhetoric of classlessness, remain primarily the defense organizations in the political field of business and property. What they really 'aggregate' are the different interests of the dominant classes . . . [But] these interests require ideological clothing suitable for political competition in the age of mass politics; one of the special functions of conservative political parties is to provide that necessary clothing."[10] While Poulantzas sees the state apparatus as completely controlled by economic forces, Huntington considers the political system to be completely independent of national or international economic forces. Ultimately, Huntington's analysis is a sophisticated polemic in support of a strong, stable state which appears autonomous but is actually upholding rule by a potent capitalist class.

Non-Marxist and non-functionalist political sociologists who work in the Weberian tradition, like Reinhard Bendix[11] and Edward Shils,[12] have emphasized the importance of the political system in sponsoring modernization and economic development. But they have not specified what type of state organization is most likely to sponsor successful development. They have talked of a strong, effective, bureaucratized or centralized state, but they have not considered how such a state apparatus relates to other sources of power in society, especially class-based power. Thus, like Marxist and functionalist political sociologists of Third World societies, they have not looked at the relationship between the state apparatus and dominant classes as an independent variable determining the type and rate of economic development.

The autonomous state in nineteenth-century Europe was an exception—under Louis Bonaparte III and Bismarck—that compensated for a weak or conflict-ridden capitalist bourgeoisie. Most early European capitalist development occurred in a bourgeois state. But in late-developing states the class base of the state is often problematic or in flux. It cannot categorically be said that one particular class always or usually controls or even benefits from state action. Rather, the class base of the state has varied between countries and within a given nation over time. Moreover, an autonomous state apparatus is much more common in late-developing polities.

The military bureaucracy often functions as a pivotal element in society, an element guaranteeing state autonomy. This is true because military elites are more likely than civil bureaucrats to be free of ties to the economically dominant class (or classes). It is also true because force is usually needed to destroy both internal and external class alliances that block integrated economic development. Economic industrialization involves new patterns of consumption corresponding to higher levels of income owing to the depletion of natural resources and/or geographical specialization in crops. Economic development, in contrast, involves the accumulation of capital and adoption of more effective productive processes.[13] As in the Middle East, Africa, and some Asian countries, it has become the unique mission of the military sectors in Latin America to "liberate" the state from class fetters in order to promote development. Liberal parliamentary regimes controlled by landed and/or bourgeois interests often promoted capitalist development in Europe and the United States, but similar regimes have proved incapable or inept in promoting development in the Third World. A parliamentary regime allows either pre-capitalist classes or a bourgeoisie tied to external capital, or a combination of the two, to dominate the state apparatus. It may use the state to oppose development. Economic development means the demise of the landed aristocracy, but it threatens the bourgeoisie as well. The bourgeoisie finds it safer to invest in foreign security markets or overseas banks. International capitalism provides more certainty of return; the risks of national growth and, more important, of economic calamity are thus avoidable, a possibility not available to an earlier "entrepreneurial" bourgeoisie.

Historically, powerful class interests in Latin American states have preferred a weak government. Most state apparatus could not sustain any role independent of the aristocracy or the bourgeoisie, and they were tied to external capital. As a result the posture of Latin American states toward Europe and North America was weak, not simply because of economic dependency but internal political dependency as well.

The present-day incidence of military intervention and then overt military rule in one nation after another in Latin America is due to the special conditions of (1) being a part of the Third World and (2) last in the evolution of economic development. To realize the goals of a full-blooded nationalism, militarism becomes the necessary pivot of rule. This is not to say that the military is pernicious or evil, or that the Latin peoples have any peculiar pathological propensity for military rule, but that the military is a means of reaching the goals of national cohesion and integration.

An autonomous military bureaucracy can initiate measures to foster economic development, but it also must contend with existing and emerging class forces. It is the strength of these classes, and how the military bureaucrats relate to them, that determine the distinctions between Latin American nations today. There are three patterns of successful late development which illustrate the potentials and problems in different Latin American countries.

1. *State-initiated national capitalist development*: A combination of autonomous military and civil bureaucrats gain power through revolution from above. They expropriate foreign property while permitting, and carefully regulating, a small amount of foreign investment. They also break the political power of landlords and often their economic power. The state helps landlords increase their productivity or it expropriates their lands. The state bureaucrats accumulate capital through taxation of agriculture and

themselves establish investment banks and basic industries, while also stimulating private investment on the part of the national bourgeoisie. State ownership of industries is only an intermediary step, however. As the national bourgeoisie grows stronger (augmented by ex-bureaucrats), the state sells off its industrial holdings. The national bourgeoisie becomes economically dominant but remains politically subordinate to the state bureaucracy, for it is unable to present or promote its class interests as the general interests of society. As a result, the bourgeoisie dominates parties and parliaments (sometimes in alliance with the capitalist landlords or a weakened pre-capitalist landed class), but these democratic institutions remain subordinate to bureaucratic power. The costs of this mode of development are greater monopoly concentration (in a small number of banks and industries) and greater social inequality in the process of development than under earlier market capitalism. The crisis in this system occurs with the rise of an industrial working class after developmental takeoff. Autonomous bureaucrats have never successfully organized a mass political party.[14] If the bourgeoisie succeeds in mobilizing mass support through party organization, it may take over the state and thwart development by allying itself with international capital to benefit its own narrow economic interests. The military's attempt to thwart such a move may lead to a series of coups. This produces political and economic instability unless the military can stabilize its rule by integrating the masses into bureaucratic sectoral organizations (unions, peasant leagues, etc.) politically subordinate to the state bureaucracy.

2. *State-initiated dependent capitalist development*: This model has many similarities to the national capitalist model, but there are several critical differences. Here, too, an autonomous coalition of military and civil bureaucrats sponsors economic development by breaking the political power and economic dominance of the landed oligarchy. But these bureaucrats sponsor and strengthen that sector of the bourgeoisie most closely tied to international capital to the detriment of more nationalist entrepreneurs. In order to prevent complete subordination to international capital, the state plays a more sustained economic role than in nationalist capitalist development. The public sector of the economy retains an important role. The international bourgeoisie becomes economically domi-

nant, but it remains politically subordinate to the bureaucratic apparatus. As in the nationalist model, leaders of mass organizations are incorporated into the political system, while most of the population is depoliticized. Several weaknesses threaten the viability of this mode of development. The autonomous bureaucrats have found no ideology with the effectiveness of nationalism to legitimate their political power. Moreover, a real question remains as to how far such dependent development can progress. Its success may depend on an internal colonialism exploiting and subordinating other less developed countries in the area as a subimperialist power.

3. *State-directed socialist development*: Autonomous civil and military bureaucrats who gain power through mass revolution destroy the landed class and national bourgeoisie and expropriate all foreign investments. They substitute for the bourgeoisie by taking over all major economic tasks and mobilizing the working class and peasants (or one sector thereof, as in Chile under Eduardo Frei) for development without giving them either political or economic power. Even when the masses have created a socialist regime, they never controlled the development process.

Workers and peasants may have considerable political and economic initiative and control at the local level (as in China), but major economic and political decisions are made at the top without popular participation. Under state socialist development, the social and economic costs of development are shared more equitably than under capitalist development, but the probability of coercive economic and political controls is higher. The crisis in this system occurs when the national resources and market are too small adequately to support autonomous development (as in Cuba). In this case, the state bureaucracy (or part of it) may become economically and politically dependent on an external power, with some of the problems that a dependent capitalist regime exhibits.

In each of the above models, military and civil-technocratic elites have more political and economic power than their counterparts in earlier developed states. In all three models the mass of the population (urban and rural) pays for development as it did in earlier eras. The important questions to ask are (1) which of these modes of development is most likely to be successful today in Latin America; (2) under what conditions is one as opposed to another type likely to occur; and (3) what are the common and distinctive costs of each model? To begin to answer these questions we must turn to more concrete analysis of specific Latin American countries.

Whatever their sharp disagreements on the relationship of the economy to the polity, Marxism and functionalism share a belief that the state is essentially capitalist in character. For my part, the very power of the modern state is defined not by the power of the bourgeoisie under capitalism or of the proletariat under communism but by the distance the modern state puts between political power and economic class. The huge, expanding bureaucratic army of messengers receiving state paychecks, from the army officers on the one hand to pacifist educators on the other, owe their primary and central allegiance to the state as such. In this sense, as I noted some time ago, Hegel, rather than Marx, has carried the day. The state regulates, mediates, and adjudicates the claims of social classes, but its essential force derives from limiting any form of class power that does violence to national interests as the state perceives those interests.

Instead of withering away the state under socialism has become more powerful than it has ever been. As Solzhenitsyn has made clear, labor camps were filled with workers opposed to the Soviet state, not to the socialist dream.[15] The ten million incarcerated in the Stalin era and the liquidation of another eight million are a durable testimony to the ability of the state to remain flexible enough to serve diverse class interests. The execution of enormous numbers of people on a grand scale indicates that state bureaucracy or state authority itself developed in a socialist world very much the way it did in the capitalist world, with fewer constraints and circumspection because of the absence of parliamentary democracy, but with the same sense of autonomous development. The state did not wither away through either the beneficence of the bourgeoisie in the West or through the egalitarianism of the proletariat in the East. The state grew exponentially while the older class system tended to stagnate. The organs of state terror, as well as the instruments of administrative efficiency, developed a huge cadre of personnel with roots only in the bureaucracy, a cadre thoroughly disabused and cleansed of class claims or privileges. Thus, in both the First and Second Worlds, the state did not "wither away"; rather classes showed signs of decimation or stagnation.

In Latin America the same process of state power evolved, but at a later date and under circumstances of relatively entrenched classes supported by foreign interests. The state in Latin America has developed powerful tendencies toward bureaucratization and, one might add, even premature centralization procedures. It has developed sensitivities to national claims, but it has done so belatedly in the face of a powerful class preference for a weak government rather than a bridled bourgeoisie. Latin American states cannot perform to their maximum potential because their classes have already developed and consolidated their power in opposition to the state. Thus, a situation has arisen throughout the continent—in Mexico, Brazil, Peru, Argentina, and Uruguay—in which the bourgeoisie grew rapidly, albeit dependently in alliance with foreign capital. A bourgeois state emerged rather than an autonomous state. The working classes grew. Trade union movements in Argentina, for example, are sizable, proportionately larger even than they are in North America, which means a state beholden to proletarian power and hence not in a position to limit working-class demands. The peasantry developed its own advanced forms of revolutionary behavior at certain times, and this too has hampered an autonomous state authority from controlling the peasantry. The Latin American state, for a longer time than states in Western Europe, tended to stagnate. For many years it could not maintain any role independent of the bourgeoisie, independent of the proletariat, independent of the aristocracy, or independent of the peasantry. Therefore, Latin America's posture toward Europe and North America was weak, not as a result of simple economic dependency but as a consequence of political dependency on an economic elite sustained by external sources.

Conditions in Latin America are different from those in North America or Europe, not simply because the area is abstractly backward but because the state is not able to fulfill its unique functions of modernization and nationalization. That is to say, these Latin American states were unable to effect what were the requisites of development elsewhere. They could not develop national postures. Instead, Latin America has developed class postures. The development of international alliances among the bourgeoisie of Latin America could be described in Marxian terms as typical behavior in a bourgeois state. But to the degree that they have such external relations, the bourgeois states of Latin America have been unable to create a climate of real independence of foreign domination.

The unique mission of the military sectors of Latin America has been to complete the "liberation" of the state from its bourgeois clients. The military sector of the bureaucracy becomes the national sector, the epitome of the state itself. Thus, the rise of military intervention and then overt military rule in one nation after another in Latin America is a function of the general law of statism: the increase of centralized power at the expense of separatist class, racial, and religious interests. This takes place under Latin America's special position of being part of the Third World and relatively late in the history of modernization and industrialization. It is thus like North America in formal legal structures, but like Africa in its patterns of economic evolution.

The military provides the means of reaching the goal of national cohesion and integration. Under such terms, the class factor vis-à-vis the national factor becomes a less important aspect of identity. In Europe in 1914 people did not identify themselves as members of an international proletariat. French workers were quite ready to kill Germans in the name of French sovereignty; and German workers were quite ready to kill any member of the French working class in the name of the Kaiser. A similar situation prevailed in Latin America in the post-World War II period. Strong nationalist feelings developed belatedly: the ferocious independence of Argentina from Brazil, and the Caribbean area from Mexico. In the Caribbean, for example, "imperialism" may refer to actions of the Mexicans or Venezuelans rather than to those of the United States.

Statism developed at a later point in Latin America and, therefore, at a point in time when social classes were relatively well advanced. Militarism in Latin America cannot be viewed simply in terms of the norm of illegitimacy, as an effort to rotate power, or as a surrogate for politics of a democratic order, but as a mechanism to harness these social classes to statist goals. Such attributes are real enough, but they remain special elements within the military state. The norm of illegitimacy, while authentic, is nothing other than a special case of the general law of bureaucratization in Latin America as a whole.

The military in Latin America guarantees state autonomy; it creates the same conditions of state power that obtain in older industrial powers. The military creates preconditions for growth, even though at times the price is high in parliamentary and democratic terms. This is entirely consonant with the history of state power in Europe. For whether it develops through the party apparatus, economic classes, or social pressures, the state has come to function as a coercive mechanism which deals severely with those who would limit its growth on behalf of class, racial, or ethnic parochialism of any kind.

That the military fulfills the role of state purification in Latin America is, however, exceptional in terms of classical assumptions that the military is dependent upon other economic forces. The military establishment, after all, does not plow the soil, does not work in factories, does not manage business enterprises, and does not take commercial risks. As a result, it must make bargains and enter agreements with those other social classes which provide the necessary preconditions for a society to exist, not to mention grow. The unique role of the bureaucracy as a whole is to manage state power and the concomitant apparatus of coercion. At the same time it remains dependent on the producing classes for taxes and wealth. The military in this sense is the vanguard of the bureaucracy of Latin America. It does not replace social or economic classes, nor can it solve international monetary crises, but by virtue of its control of the mechanism of coercion, and hence the power of persuasion, it can adjudicate the claims and interests of classes on behalf of what is presumed to further national development.

This situation can be framed in social contract terms: classes exchange a certain portion of their power for the protection and care provided by the state. Through the machinery of taxation, such classes pay for police functions as well as military functions. What any ruling economic class wants in this exchange is the ability to maintain its class position without "trouble" from other classes. The Latin American military, in relation to mass impulses toward democracy, performs essentially this kind of law-and-order role. However, all the Devil's bargains have their prices. The bourgeoisie, in nations like Cuba, often gets less than it bargains for. The military does render the services contracted for, but the asking price is not only a monopoly on powers of coercion but a limitation on the powers of the classes that installed military rule in the first place. The delicate balancing of forces between classes and bureaucracies largely defines the Latin American system.

In a nation such as Brazil, the bargain is struck with a relatively feeble, even corrupt, bourgeoisie that is unwilling to pay taxes, to invest in the Brazilian national recovery, and to take the risks that captains of industry in earlier centuries thought normal and even necessary. It is important to note that in a nation such as Brazil one does not have a swashbuckling entrepreneurial state but rather a bourgeoisie that, given its choice, would prefer to invest in the New York Stock Exchange or in Swiss banks. This is not to single out the Brazilian bourgeoisie. On the contrary: for the most part, they have been far more vigorous than the bourgeoisie in other Latin American countries. But the Brazilian bourgeoisie was never able to develop the sort of monopoly of power necessary to rule. Instead, they were involved in an endless series of class compromises and group accommodations with other social sectors. The Brazilian bourgeoisie was large enough so that no social force could rule without its support, but feeble enough so that its will could never become identical with the monopoly of state power as such.

Under these circumstances, the military of Brazil struck the bargain which, in 1964, permitted the bourgeoisie to survive as a class, but only on condition that it become fiscally responsible and economically honest. Unlike the European middle classes, honesty was not just a necessary characteristic of the bourgeoisie but a national prerequisite for an advanced export economy. Brazil fast became a moralistic nation in which the military imposed on the bourgeoisie the same kind of restrictions that in an earlier era were imposed on economic classes in the United States and Europe. "Brazil's military rulers have followed policies that have deprived most political actors of their autonomy, narrowed the arena of 'permissible' political choice, and sought to eliminate all alternative system outcomes except self-perpetuation in power."[16]

The political agreement, once struck, permitted the bourgeoisie not only to survive but to grow tremendously. But this meant an abandonment of political democracy, and the direct imposition of military rule for a high gross national product. In an earlier political democratic phase, Brazil showed much less growth in the economic sector. Indeed, limited growth approaching stagnation between 1960 and 1964 spelled the doom of civilian rule in Brazil. After power was seized by the military, the gross national product rose to annual growth rates of between 11 percent and 12 percent, where it leveled off in 1974. Thus, we see that military rule, by its coercive impositions on all social classes—indeed, what has been termed a "pure colonial fascist archetype"[17]—began to spark an economic takeoff very much the way Europe managed economic gains with civilian rule.

The military alliance with the bourgeoisie in Brazil, Chile, and elsewhere in Latin America is not illustrative of a new social equity. It is not a system based upon the distribution of goods in equal terms or fair shares; rather, it entails an impetus to internal growth by freezing the status quo in stratification terms.[18] The elites reap most of the profitability out of this system, while residual profits slowly filter down to the base of the society. The bourgeoisie is the most advantaged hierarchically. To a lesser extent, the factory proletariat, because of capital-intensive modes of production, is somewhat advantaged, while the peasant masses are hardly affected by this new wealth of the Brazilian miracle,[19] The conclusion this leads to is that the institutionalization of an authoritarian system by the Brazilian military will become successful in terms of a turn toward nationalist, anti-imperialist, and even anti-American policies.[20] Of course, the success of this approach depends on the prior achievements of the Brazilian economy, and that includes its own imperial pretensions. It has even been suggested that in Brazil this "national authoritarianism" takes place within rather than outside the imperialist domain.[21] But the push of continental leadership may more readily lead to counter-imperialism than sub-imperialism. In any event, the existence of a national military underwriting a nationalist authoritarian polity is scarcely deniable by serious analysts of Brazil.

In Argentina the situation is seemingly different. In that country the military had ruled either overtly or covertly since 1930; but the military in Argentina has been increasingly isolated from the middle classes and basically has ruled in isolation from the class system as such. As a result, the military, especially in its post-Perón period, from 1956 to 1973, was unable to "civilianize" itself sufficiently. Thus, it could not become allied with the bureaucracy as a whole, and was unable to democratize itself, to enlist the support of the working class, to gain a monopoly on the sources of coercion and terror, and, therefore, unable to assist the bourgeoisie in its own survival and growth. As a result, the Argentine military, during the interregnum period, illustrates an earlier pattern of covert military domination rather than the newer pattern of overt military control.

The second Perón era of 1973–74, short-lived as it was, brought about the reconciliation of the military and the *Justicialista* movement and, in that sense, with the state apparatus as such. It is a curious fact that a purely class victory or class compromise remains elusive. "Arrangements" are made between classes and elites: in Argentina between the party representing the masses of *descamisados* and the military representing the critical, urban, exporting bourgeoisie. In Cuba "arrangements" are between the military guerrillas who made the revolution and a steadily emerging urban communist party apparatus, both of which were underwritten by a peasant proletariat. In Brazil, and with some variations in nations like Peru, "arrangements" are made between military elites who provide moral governance and a bourgeois class which provides economic rationality.[22] Thus, Latin America does not fit into the classical European model of class compromise or class struggle but represents an agreement along lines of a ruling bureaucratic elite (usually but not always the military) and a powerful economic class (usually but not always the bourgeoisie), neither of which can long survive intact without the other and neither of which can quite come to terms with the tasks of political leadership.

The reason for neo-Peronist strength, which permitted a return of Peronism after Perón, is that the bourgeoisie for a time set the conditions for continued military perseverence rather than the other way around. The end result was similar—the military possessed a monopoly of power and coercion in exchange for their stimulation of capitalist production. To establish this parity in Argentina, however, the power of the industrial classes had to be built up, while the power of the military was diminished by the same proportion.

This helps to explain the existence of both powerful Right and Left tendencies in neo-Peronism: both the trade union movement and the bourgeoisie had to be admitted to the source of power by the revitalized leadership as the essential mechanism for curbing military authority. The important point is that, despite the cashiering of this anti-Peronist faction, the military has not dissolved civilian politics in Argentina. Its role as national guardian was in no way impinged upon by the return of Perón to power; indeed, since the demise of Juan Perón this guardianship role has substantially increased. The historical working out of the arrangement between military and polity is different in countries with such distinct histories as Argentina or Brazil, or for that matter, Peru and Chile. But at present developments in the structural aspects of military systems are limited with expanding class economies throughout Latin America.

Recent developments in Peru might well be prototypical for the pattern of civilization in the Third World, or at least those sectors of the Third World with a high degree of class differentiation and well-developed political institutions. The military came to direct power in the late 1960s through a then typical coup led by Juan Velasco Alvarado. This in turn provided the basis for national consolidation, expropriation of key foreign holdings, and administrative rationality. But once the military got beyond expropriating and redistributing land, the more complex problems of foreign debt and high inflation, and, equally, of high unemployment became tasks more easily dealt with by civilian regimes. What followed was a transitional military regime led by Francisco Morales Bermúdez, who paved the way for a re-establishment of multiple party forms and electoral norms.

The stunning pattern of the 1980s, witnessing the re-emergence to power of Fernando Belaúnde Terry, represents an effort to solve serious internal economic problems, and also an expression of long-term political trends, in which the militarization process serves to institutionalize the specifically Third World characteristics of a nation. This process then stamps the country in a fundamental military way, as indeed the economy stamps the First World and the polity the Second World. It is significant, for example, that a new Peruvian leader expected to have "very fruitful relations with the armed forces"; and was interested in establishing a "broadly based government of national concord." [23] If this is a pattern rather than an aberration, the military origins of many Third World systems, at least present-day systems, does not preclude the formation of democratic systems; an evolutionary process of considerable significance not only for the structure of the Third World, but for the tranquillity of international relations as a whole.

In Cuba, which one might call the acid test of this hypothesis, the same kind of relationship is established between the military and the proletariat as exists between the military and the bourgeoisie in the capitalist countries of Latin America. There is no denying the communist character of Cuba, any more than one can deny the capitalist character of Brazil or Argentina, but the definition of the economy of these countries in no way precludes the military control of the system of state authority. As the Cuban military increasingly reveals its pre-eminence in the party—if not over the party—the same kind of Devil's bargain as exists in capitalist countries, in which the military guarantees tranquillity in exchange for high economic productivity, becomes the enshrined pattern. The same structural arrangements and the same economic consequences obtain under militarized "socialism" as under militarized "capitalism." This is because, as Third World nations, military patterns of domination are central.

In the experimental phase of the economy, in which peasantry and proletariat ostensibly ruled through the party, Cuba had the same zero growth rate as was to be found in Argentina and Brazil under civilian rule. Only with the adaptation of a military system did the Cuban economy show any signs of an economic turnaround. In 1973 and 1974 the rate of growth of the gross national product of Cuba increased by 5 percent and 6 percent respectively. This growth occurred in a nation which in its pre-military period exhibited virtually negative growth rates. The drop back to zero growth was more a consequence of Soviet needs than Cuban potential. Once again, the function of the military is not simply to monopolize coercion but to assure development. The Cuban military is far less concerned with ideological questions than with production norms. In Cuba, as elsewhere in Latin American, the size of the military sector is determined in large measure by the surplus wealth gained from the developmental process, not by the surplus rhetoric generated by ideological disputation. Thus, the Cuban case is a special variety of the growing militarization of the Third World.

The Mexican model represents another special situation. Mexico had a mass peasant revolution in the name of concepts of nationalist *indigenismo,* and only slowly, over time, did it develop a capitalist system. A special irony of the Mexican situation is that a Leninist theory of the state, a one-party system, performed the essential requisites of coercion, and hence not only permitted but encouraged the bourgeoisie, between 1920 and 1940 to grow in parallel fashion with other classes and sectors. The state encompassed both the bourgeoisie and the military and achieved the goal of state autonomy without overt military rule. Such civilian rule did not automatically yield democracy but rather a refined form of Third World Bonapartism.

The Mexican political structure was able to contain the military by providing it with a center of power. The military factor was built into the state system, and hence it did not have to overthrow what did not exist to begin with: a multiparty parliamentary democracy.[24] Even so, it is instructive to note how powerful the Mexican military has become over recent years, despite the fact that the military was contained and restrained by the rise of a single-party state system. Over time the military has emerged, along with the bourgeoisie itself, as one of the two critical factors in Mexican state power. The outcome of the Mexican Revolution has been neither peasant power nor proletarian power, but a special kind of coalition of the bourgeoisie and military under the rubric of the party system that obtained elsewhere in Latin America. The evolution of this ruling-class military coalition was distinct in Mexico from elsewhere in Latin America. However, the outcome exhibits similarities with the continent as a whole, and thus underscores the position herein taken concerning the militarization of the continent as a function of the need for autonomous state power, whether in the name of state capitalism or state socialism.

In Latin America problems of state power shift with the bureaucratization of the administrative process. Issues relating to production are subordinated to problems of allocation. This particular dimension in Latin America often results in a system that allocates imported goods rather than promotes domestically manufactured goods. The role of the state becomes even more important than in nations in which capitalism establishes its own network of marketing, distribution, and allocation. For example, in Latin America the extent to which one engages in real estate speculation is directly tied to one's level of advancement in the state bureaucracy. Hence, problems of commodity and realty allocation are from the outset political. Both are largely in the hands of a state apparatus rather than an economic class. As part of the bureaucratic apparatus the military performs the allocational functions of the state, whereas capitalism in North America or Western Europe maintains autonomy in the economic marketplace.

From the outset, the Institutional Revolutionary Party (PRI) chose to be ideologically pragmatic and vague in order to accommodate various groups. Groups of different persuasions were offered rewards and concessions in return for loyalty to the party and the regime. However, it was made clear that the party could not tolerate strong regional centers of power that were outside of or not allied with the party. If cooptation failed, strong-arm methods were used. Many regional *caudillos* were eliminated on orders from the regime. The pendulum of cooptation and repression of dissidents gave the party a dominant role over the military. But since the mechanisms of repression were themselves militaristic, the armed forces grew and prospered in Mexico in alliance and conjunction with the one-party state. The very absence of a multiparty parliamentary system made possible the realization of armed control goals without the customary side effects of overt military rule. The Mexican armed forces are one of the best equipped and most professional military groups in Latin America, and got to that point by their firm adherence to the single-party state.[25]

Under these conditions the military provides a stabilizing point for the bourgeoisie, a way of controlling the flow of goods, monies, and services in order to permit the national economy to flower. Somewhat analogously, in a country with no bourgeoisie such as Cuba, the military permits the party apparatus the same latitude and the same operational potential for autonomy. In short, the military guarantees, as only it can, a place for a ruling class that otherwise would be easily overwhelmed by imported goods and exported capital.

For a long while there was a strong belief that the military of

Latin America simply performed internal police functions. Left to Right agreed that these are not professional military personnel but political military personnel. A further assumption was that their main obligation and chief responsibility involved internal repression rather than the management of international conflict. This is only partially true. It is dangerous to overemphasize the dichotomy between a politicized military and a professionalized military. Increasingly, the role of the military is directed toward strengthening national claims vis-à-vis sub-national sectional claims on the one hand and international imperial claims on the other. The military has long been considered a factor in the political growth of new states.[26] But the conventional literature argues with rather than interprets reality. Military rule is confused with totalitarianism, and the overall superiority of political democracy in gaining the desired ends of change is argued. Such an approach presumes that modernization rather than industrialization is the chief goal of development. This premise requires the most critical scrutiny, since the needs of industrialization, rather than those of modernization, seem to call into play the military pivot. Beyond that, one can expect warfare to take place between various Latin American nations in which the military emerges as a professional cadre and not simply as an internal repressive cadre. One can easily postulate conflicts between Central American nations, or between Peru and Ecuador where border conflicts are constant and real, or between Argentina and Brazil over the fate of smaller nations in the area such as Uruguay and Paraguay. Indeed, the themes of nineteenth-century revolutionaries have not died down nearly as much as the linguistic hegemony of Latin America would lead one to suspect.

Several military establishments are emerging in Latin America that have a real capacity for foreign action, albeit limited to the continent itself. The inherited belief that the military serves only to repress the popular will (a system maintenance role) has limited application. The military may eventually express the popular will by conducting national warfare (a system transformation role).[27]

Let us turn now to the relationship between the United States and its military establishment on the one hand and Latin America and its military establishments on the other. The overwhelming impression from the statistical data is that the development decade of the 1970s, like the earlier periods, generally resulted in trade and aid stagnation. The relative conditions of Latin America vis-à-vis the United States have either not improved or have, in real terms, deteriorated even further from what they were earlier in the century. But this aid factor is not indicative of the real growth of the military despite this economic period of stagnation.

An important aspect of the aid-trade situation is that in terms of hardware the United States assistance to the military rarely amounted to more than 2 percent of the gross national product of various Latin American countries. What has increased considerably, indeed tremendously, is the portion of the gross national product that has been allocated toward the purchase and outright acquisition of sophisticated and advanced military hardware. According to the most recent figures, Latin America, as the largest buying area within the Third World, has purchased a record $1 billion worth of hardware from the advanced countries over the last five years.

From the point of view of military equipment, the United States input into the area has been enough to maintain a major role in the armed forces of the hemisphere but not sufficient to change the structure of relationships between the United States and Latin America; certainly not enough to tilt those relationships so they would become more favorable to United States interests over and against national interests. This is not to deny that a huge factor in Latin American military strategy and tactics continues to be the response of the United States. It is to point out that this United States role has remained relatively constant throughout the last thirty-five years and cannot, in and of itself, explain the movement toward either the militarization or nationalization of South America. Indeed, one might very well argue that, as the Latin American military gains relative autonomy from the United States military, it also performs its new nationalist role.

During the last decade the military of Latin America has gradually transferred its allegiance from the imperial sphere to the national sphere. Insofar as the military has grown strong through equipment purchases, it has severed ties of dependency and has been able to spread its purchases of hardware to all the major powers of the world. There would be more spread were it not for the technological limitations imposed by non-interchangeable parts and the differing weights and measurements used in manufacturing military goods. The military has, in fact, become nationalized. Its concentration of force enables it to rule within the national domain. For the first time, American political interests have become aware of the fact that democratic civilian regimes may provide a better buffer for United States interests than undemocratic military regimes, especially insofar as these military regimes move in step with the imperial fortunes of their own democratic regimes, as is the case in Brazil, Mexico, and Argentina, or with Sovet interests, as is the case with Cuba.

One might say that the current United States posture represents a newfound reluctance to place faith in a hardware solution. Bolstering the military sector in the name of civic action and civilian defense has clearly run its course—and with scant benefit for the United States. The rise of militarism in Latin America has not prevented the takeover of United States properties. Even nations friendly to the United States, such as Venezuela, show a powerful impulse to national solutions (as in the case of oil reserves) at the expense of any presumed international or South American hegemony.

The new military and the new Left of Latin America jointly sponsor a kind of nationalism which presumably is indifferent to the United States and its interests. Even the most reactionary regimes, such as post-Allende Chile, clearly intend to weaken further the fortunes of overseas imperialism at the expense of their own national bourgeoisie. The military, in fact, has become the national nerve center in every country in Latin America, even where civilian rule ostensibly exists. Assuming the benefits to be automatic and discounting the liabilities of doing business with the United States is no longer pre-ordained. Currently the United States recognizes that it is dangerous to equate long-run secular interests with entrepreneurial policies. As a result, American political elites recognize that the military is allied to United States interests, but only in a dubious and shaky manner. The military may prevent the spread of national revolution, but only if there is a serious reallocation of North American priorities to the greater benefit of existing national classes.

The new conditions of military domination of the state and

state independence of the bourgeoisie have led to a tactical reassessment of the situation by both United States and Latin American leaders. Henry Kissinger defined the need for a "new dialogue" not as a concession but as a collective necessity. But the terms of this new dialogue are under the most vigorous dispute and scrutiny. Luis Echeverría, past president of Mexico, in responding to the new dynamics in hemispheric relations, indicates that militarism and statism do indeed require North American *concessions* and not simply collective *integrations*: "We need an overhaul of international economic and political relations. The blame for the problems of the world rests primarily with the big powers. . . . Latin America supports price increases for oil and other raw materials. Indeed, the oil crisis has sounded the end of an era based on injustices of consumer societies."[28]

The "anti-American" formula is not intended to overthrow bourgeois power or to institute an age of democracy, but, rather, to employ military dominion over state power so as to minimize foreign control and maximize national control. This is the formula for the realization of economic autonomy followed to the letter by the older capitalist powers in their own earlier period of expansion and growth. History imposes new forms on developmental approaches in regions which are beyond colonial status and which have had their integrating revolutions. With our great material and technical resources we were often tempted to do for others what we thought was best for them. That attitude no longer shapes the relationship of superordinate to subordinate powers.[29] The United States has come to recognize that a revolution has taken place in Latin America. Industrialization and modern communication have transformed economic and social life. This is why a dialogue with the Third World is not a concession by the United States but a necessity in a universe where real shifts of power within the capitalist economic sphere have taken place.

The Latin American military, in relation to the United States, supervises the process of development in order to maximize its own national advancement. The military seeks to do so in ways that will maintain internal stability and the growth of the current ruling classes. One should not be under the illusion that military determination will dissolve under the pressure of Western-style liberalism. The military pivot derives its strength from an authoritarian model and is not about to change this political thrust, any more than it will voluntarily alter its sense of class purpose.

We are witnessing a new form of coalition, one that is far more favorable to "peripheral" than to "cosmopolitan" interests between Latin America and North America. In this new statism the major powers will continue to remain central to any determination of the Third World. On the other hand, the underdeveloped powers realize that they have certain leverage at the level of military mobilization from within and mineral pre-eminence from without to prevent any return to the older style of colonialism. The new military nationalism no longer permits a situation in which profits are sent abroad and high interest rates are absorbed by the peripheral sectors. The Third World has come to employ the military, not as an intermediary sector, not as a force for Machiavellian interests at the political level, but as the main pivot in the administrative process of development. The process is not carried out by civilian authorities in North America and Western Europe, but by the military sector of less developed areas. In this manner the emergence of a nationalist military sector is but one form of the international bureaucratization of elites. The new nationalism emerging throughout Latin America stimulates modernization in a way that will permit competition with the older imperial powers of the East. The prospects for success of this military style of development will depend on a host of imponderables, not the least of which are the skills of the armed forces themselves in rethinking no less than reorganizing. Such long-run estimates of the survival potential of the military in Latin America are outside the purview of this chapter.

Only now, as we approach the twenty-first century, are nations such as Brazil, Argentina, and Mexico realizing the potential said to be theirs at the beginning of the twentieth century. Whatever the explanations—local, national, or international, cultural or sociological—prior to its manifest militarization, Latin America was in a condition of critical economic stagnation that can only be described as the transformation and inversion of private-sector capitalism into public-sector capitalism. The implausibility, even the impossibility, of such economic retrogression has finally been faced by Latin America. The military has carried forth this new consciousness in an effort to reverse the trends toward backwardness and to enter the modern world; to pay the heavy costs of social change and be sure that others pay it, willingly and otherwise; and to pay even heavier costs in political liberties or at least be sure that others pay it. This military theory of politics may seem pessimistic and certainly provides a less than exuberant sense of what the Third World is about. It does provide a ray of hope in that the fundamental loyalties of Latin America to the development process have finally replaced the illusions of North America as a basis for further analysis and political practice.[30]

9

Legitimacy and Illegitimacy in Third World Regimes

In the study of legitimacy we are the children of Max Weber; just as in the study of revolution we are the offspring of Karl Marx. If there is any truth to the statement that Weber is the "bourgeois Marx," it shows up in the analysis of how social systems sustain themselves over time through a combination or permutation of traditional, bureaucratic, or charismatic modalities. That Weber himself appreciated the complexities inherent in the concept of legitimacy is indicated by his allusion to the special circumstances surrounding the Papacy. He saw such church forms of legitimacy as a function of the "charisma of office." Leadership potency is derived from the institutional properties of a world historic church rather than the personal qualities of any given Pope.

If charisma of office is a distinct source of legitimacy, then a present-day secular variant may be called charisma of party. In *Three Worlds of Development*, I emphasized that such a phenomenon seemed to have arisen throughout the Third World and non-aligned nations.[1] In Egypt under Nasser, India under Nehru, and Yugoslavia under Tito, the major single-party units survived intact after the demise of an originating leadership. Indeed, it is a tribute to the innovative potential of Third World political systems that they have managed to address problems of succession and leadership in such a creative manner.

Since the first edition of *Three Worlds of Development* it has become apparent that the notion of party charisma cannot be said to be a general law of political development. When institutions have been durable and powerful in their own right, such a concept has great operational worth, for example, in describing the leadership of the Roman Catholic church or in certain socialist or Third World countries. However, in other nations, from Ghana to Brazil, the party apparatus itself was subject to extreme pressures, and with the fall of charismatic figures like Nkrumah or Goulart, party dissolution also occurred. This happened to a greater or lesser degree, to be sure—sometimes the party was absorbed into new political formations—yet it was no longer possible to subsume all or most Third World or non-aligned powers under the rubric of party charisma.

It became evident by the late 1960s that military might in the Third World had become pre-eminent. Further, military management of the political system was not based on anything remotely resembling Rousseau's social contract, but was simply a rearrangement of elites lacking either electoral or popular support. These new structural configurations in the Third World were, and remain, so profound and so entrenched that it was no longer feasible to speak of a "crisis" in legitimacy. Rather, it became necessary to explore what the new configurations of power were, and how they helped form a new normative framework, quite unlike the configurations that characterize Western and socialist societies alike. To these new configurations affixed the term norm of illegitimacy. The norm of illegitimacy concept represented an effort to replace "party charisma" with a more embracing concept, one that takes into account the phenomena of party, class, and state under praetorian conditions.

The norm of illegitimacy was originally intended to explain what appeared to be the aberrant and yet permanent political climate of crisis throughout Latin America.[2] It is clear that the model is applicable to many other parts of the Third World as well. The so-called crisis of legitimacy has now clearly invaded intellectual discourse in the First and Second Worlds as well. We have an expanding literature of growing significance on the breakdown of legitimacy in the West,[3] and a similar sort of doomsday literature covering contemporary communism.[4]

Militarized Politics

What is occurring is less a crisis of legitimacy than the emergence of new forms of institutionalizing power. Seen in this way, problems of legitimation do not necessarily evaporate, but they do take on a more specialized meaning. Problems of legitimacy are seen in long-standing terms rather than in the conventional mode of crisis thinking about how to maintain legitimate authority. Militarized politics acquires a normative standing equal to if not greater than the older capitalist and socialist orders.

The crisis in legitimation has common roots, yet different forms of expression, in the three worlds, given the structural differentiation of their political systems. In Western capitalism and democracy, those societies most characterized by parliamentary systems, the problem has been that within a mass society there has been a stretching out of representational government. When a constituency is small, the relationships between representatives and those who are represented are direct and organic. But as the size of parliamentary governments is held constant, and citizen number continues to grow, the ratios change. Representatives of ever-larger constituencies are nearly as anonymous, remote, and removed from awareness of popular consensus as executive officers or even appointed officials. A feeling of non-choice develops, of helplessness before historic events presumably beyond individual control. Such feelings generate at least a short-term falloff of interest in political affairs, conventionally labeled by political scientists as "political alienation" in a democracy.[5] Whatever labels are attached, there is widespread agreement that the sense of distance between those who rule and those who are ruled is in large part the source of a legitimacy crisis in the West.

In the Second World, that of socialist economies and political authoritarianism, the problem of legitimacy shifts. In nations where an independent communist movement seized power, the basis of legitimacy is not electoral so much as based on a doctrine of direct popular participation, variously called the dictatorship of the proletariat, socialism of the whole people, the rights of the masses, and not infrequently involving a charismatic leader.[6] In socialist countries, nearly 100 percent electoral turnout on behalf of a single party is characteristic, and parliamentary government is reduced to rubber-stamping executive authority. Different political structures yield similar outcomes: feelings of non-choice and resulting falloff of popular interest are typical. We need only witness the constant efforts of the Soviet government to stimulate participation to realize how serious apathy and alienation become, and how ultimately they create a crisis in legitimate authority.

One should not confuse political legitimacy with national hegemony. Neither the American nor the Russian Revolution, each a classic model, could boast hegemonic rule by the *de jure* procedures of claiming legitimacy for a revolutionary elite or party. In the American Revolution, claims of legitimacy for the "first new nation" divided the colonists and united the loyalists, most of whom settled in Canada. Leslie F. S. Upton put the matter thus:

> The American Revolution was a civil war between British subjects and particularly between British Americans. The Loyalists were the losers. They were the first "unAmerican" Americans, interrogated by committees, condemned without judge or jury, smeared in the press, expelled from their jobs, and deprived of their possessions. The American mind can tolerate any crime but that of failure: the grand sweep of American history covered the Loyalists in oblivion.[7]

And the Russian Revolution revealed deep cleavages between Bolshevik legitimists and social democratic hegemonists. As David Lane notes, the immediate consequence of the Soviet state was civil war: "From 1918 to 1921, the country was enveloped in the chaos of civil war and foreign intervention: it was a period known as 'War Communism.' For much of the time the Bolsheviks ruled only one-seventh of Russian territory, mainly the Great Russian areas of the country, the remainder being occupied by other governments."[8] Even at points of maximum revolutionary fervor, legitimacy was never quite holistic.

The success of a revolution is itself a phase of legitimating a regime. Institutionalization follows, which operationally refers to the emergence of hegemonic rule, or at least a broad-based consensus between the holders of power. Creating an institutional framework is often as difficult and exacting as leading a successful revolution. Indeed, it has become an iron law of socialist revolutionary politics that the cadre formed in making a revolution is inevitably displaced by one formed to carry through its stated tasks. In this postrevolutionary process very few of the original "old revolutionists" survive. But quite beyond legitimating and institutionalizing phases is what might be termed the "highest state" of revolution-making: solving the riddle of succession.

The crucial point in a revolution comes when there is a need to transfer power from one person or group to another person or group; and to do so without destroying the entire governing and bureaucratic networks. The serious problems that arose in the transference of power after the death of Lenin in the Soviet Union and of Mao in the People's Republic of China are a clear indicator of just how difficult succession can be in non-democratic regimes, even beyond the question of what to do after the death of the original revolutionary leader.

There is clearly no uniform outcome to any succession crisis. For most parts of the Third World, the original "charismatic" leader is followed by a developmentally oriented military. However, as in Nigeria and now Brazil, this may in turn be followed by a powerful tendency toward civilianization. In the Second World, where elite domination accompanies revolution, single-party control remains firm, although the military factions begin to exercise a great deal more power, as they did in China. In the First World, where a market economy persists, the death of the leader is often followed by the restoration of older multiple parties and the formation of new ones. The cases of Portugal and Spain in the 1970s indicate the persistence of democratic norms even after long periods of authoritarian forms. It is helpful to note the three phases in a postrevolutionary morphology: legitimation, institutionalization, and succession. Differences in any of these phases, not to mention the unique characteristics of national cultures, influence different solutions to the problem of legitimacy.

When viewed against a historical background, urging older forms of legitimation upon the Third World is not particularly effective. It might well be argued that military politics in the Third World have resolved the crisis of legitimacy by turning the political structure upside down: by institutionalizing forms of illegitimate authority that are derived neither from the electoral process nor the direct will of the people, but from a presumed national need best understood by those who lead, and best implemented by those who are led. Faced with monumental problems of modernization at the economic level, mobilization at the political level, and militarization at the organizational level, the Third World has moved toward a style of governance that must be understood on its own terms. The arrogant presumption that such Third World forms are only a stage in an evolutionary scale that is still considerably behind both Western or socialist forms of government misses the point. Ironically, legitimacy, like illegitimacy, is now a problem for the First and Second Worlds as well as the Third World.

The Structure of Illegitimacy

The norm of illegitimacy entails ten central propositions; elsewhere I have discussed these at length. The following is a summary of the central issues in the concept:

1. The norm of illegitimacy arises when structural requisites for legitimate authority are absent. When authority can be institutionalized neither through the mechanism of class/mass, nor through law/authority, then illegitimacy exists.

2. The norm of illegitimacy serves the needs of both the political/military order that gives direction to national policy, and an international/multinational order that delimits the direction of any nation in the world economy.

3. The norm of illegitimacy is intended to produce mass political mobilization, and at the same time to reduce unsponsored violence or acts of terror. Illegitimacy becomes an instrument to prevent rather than to stimulate untrammeled social change.

4. The norm of illegitimacy is underwritten by a combination of national and international elites which view a circulation of government power among them as beneficial to larger global interests.

5. The norm of illegitimacy is usually enforced by military or quasi-military forces, sometimes overturning civilian regimes, and at other times overturning military regimes; but always circulating elites while minimizing the risk of civilian authority in the Third World.

6. The norm of illegitimacy distributes political authority without providing a mechanism for the distribution of economic resources. Since no government undergoing constant political rota-

tions at the top can legitimate itself, either electorally or through direct participation, political illegitimacy becomes institutionalized.

7. The norm of illegitimacy prevents any undue entrenchment of special sectarian interests. Hence, it overcomes structural deformities in the political system, and does so without risking social revolution or requiring systemic overhaul.

8. The norm of illegitimacy is a mechanism for managing problems rather than solving them. Its purpose is to institutionalize elite authority without the conventional calcifying effects of older civilian or military formations that are attached to race, class, or religious interests instead of state power directly.

9. The norm of illegitimacy presupposes a conflict model, just as theories of legitimation for the most part presuppose consensus models. When a consensus apparatus becomes inoperative, political-military regimes turn to a norm of illegitimacy rather than risk widespread upheaval.

10. The norm of illegitimacy permits the institutionalization of crisis as a normative political pattern; a norm of legitimacy is not only inoperative in most parts of the Third World, but increasingly subject to wide-scale pressures in the First and Second Worlds.

A considerable amount of work on illegitimacy took place outside the framework of political sociology, particularly among political anthropologists. Clifford Geertz[9] and Georges Balandier,[10] especially provided a deeper understanding of the subjective side of the processes that accompany the breakdown of legitimacy and, parenthetically, make possible the emergence of a norm of illegitimacy.

Geertz pointed out that for virtually every person in every society some attachments flow more from a sense of natural, even spiritual, affinity than from social or political interaction. He went on to note that in modern societies, the shifting of such ties to the level of political supremacy has increasingly come to be considered as pathological. An increasing percentage of national unity is maintained not by calls to blood and land, but by a vague, intermittent, and routine allegiance to a civil or military state supplemented to a greater or lesser extent by the government use of police powers and ideological exhortation. In modernizing societies where the tradition of civil authority is weak, and where technical requirements for effective welfare are poorly understood, primordial attachments demarcate autonomous political units. What finally emerges, according to Geertz, is a theory that authority comes only from the inherent coerciveness of such primordial attachments, or what I call the norm of illegitimacy.

Georges Balandier enriches this notion of the primordial or spiritual source of permanent strife by speaking of the desacralization of power. This process of desacralization of the traditional leadership of kings and chiefs was inevitable since the power of those sovereigns and chiefs was legitimated by reference to colonial governments rather than native groupings. Such authority rarely rested on any intrinsic merits of such systems of rule. Hence, the transition from patrimonial to bureaucratic authority does not take place in the Third World the way it did in either the First or Second Worlds. Because these colonial political systems were so varied, different reactions to the experiences of postcolonial transformation were inevitable. Bureaucratic desacralization did not bring about the terrible effects feared by former rulers; but it did result in types of rule that attempted to combine hierarchial and egalitarian forms rather than displace the latter by the former.

Balandier notes that the process of modernization was set in motion by the colonial adventure. But that very adventure created political distortions that led political parties, social ideologies, and military regimes to seek out forms of institutionalization quite different from older European and American societies. All human societies produce politics, and none is resistant to historical processes. The response to this universal proposition is the particular: namely, that Third World politics produces national, ethnic, racial, and religious demands that lead to a very different set of norms in the Third World than are produced in either the First or Second Worlds.

The Structure of Legitimacy

The norm of illegitimacy is deeper and more pervasive in Latin America than it was a decade ago, and in many respects Latin America's handling of political legitimacy is a prototype for the rest of the Third World. Argentina, Brazil, and Peru were joined by presumably bedrock democratic regimes such as Uruguay and

Chile in the 1970s. In part, the failure to recognize the growing militarism of Latin America, even in former democracies such as Chile and Uruguay, results from an inability to appreciate the gap between developmental impulses and economic limits, that is the extent to which these countries shared with the rest of Latin America demands for wider popular participation and greater economic growth without the wherewithal to deliver goods and services. But the failure also reflects an inbred tendency on the part of developmentalists to live in a world of labels and to underestimate profoundly the potency and power of militarism in the history of countries with democratic backgrounds such as Chile,[11] Uruguay,[12] and Peru.[13]

An area in need of far greater emphasis is the relationship between electoral politics and the norm of illegitimacy. Specifically, why should it be that elections, the foundation of constitutional legitimacy, should fail to prevent military dominion, authoritarian takeovers, and the constant rotation of self-appointed leaders? Alfred Stepan found that in Brazil the smaller a regime's margin of victory the more likely it is to be overthrown.[14] And Martin Needler broadened this observation even further by noting that "the narrowness of an electoral victory seems to impair a government's legitimacy, thus making a military coup against it more likely."[15] Yet, in North American and Western European political democracies where the system itself and not the party process provides legitimation, similarly close elections, while engendering bitterness and charges of vote tampering, seem to unite the nation and prevent militarist outcomes. While a tradition of competitive elections may itself reduce the distance between winning and losing, it also stimulates a norm of illegitimacy based on charges, often launched by military technocrats, that the electoral process is wasteful, divisive, and fruitless. And if one examines the sad roster of Latin American nations that once had vigorous presidential elections and far from routine campaigns, it is evident that few political democracies have endured. It is the quasi-democracies, the basically single-party states (Mexico, Cuba, and to a lesser extent Venezuela and Costa Rica) that have shown greater resistance to illegitimate authority than the former democracies of the area. But one might well argue that single-party states are special cases of illegitimate authority.

One must not equate military technocracy with the norm of illegitimacy. Although the norm of illegitimacy invariably requires military participation, many military systems were and remain transitional, such as those in Greece, Spain, and Portugal. It is conceivable that military rule can become institutionalized and hence develop a sense of quasi-legitimacy without constant rotation of power. This is certainly what has taken place in Brazil. The military regimes essentially institutionalize their power, regularize the political processes in order to solve the problem of succession, and in this way create political forms that stabilize rather than rotate elites in power. Brazil's more recent rulers have issued directives intended to prevent civilian opposition from obtaining power in the foreseeable future, but also to move Brazil away from the traditional personalist military dictatorships of previous years by at least giving legitimate voice to the opposition; and they incorporate the military as the only institution capable of leading the economic and social transformation of Brazil. Since time is a crucial element in legitimation, the ability of the Brazilian military to maintain itself in power far longer than either it or its opponents originally thought possible represents a move away from the norm of illegitimacy to a type of authoritarian legitimacy that was far less prevalent a decade ago.

Argentina, which provided the classic model for the norm of illegitimacy between 1955 and 1975, has also moved toward military stabilization of the political process. Under the rule of its military presidents a measure of stability has been achieved. What makes Argentina's military government different from those in Chile and Brazil, and other countries with similar military rule, is its refusal to consider military rule as permanent. It constantly defines itself as a transitional movement. This insistence on transition gives continuing substance to the norm of illegitimacy in theory as well as fact, for as soon as civilian rule is threatened, the coup within a coup occurs; and the norm of illegitimacy is preserved.

The crucial test for the norm of illegitimacy is the same for all societies: Can such a framework achieve high levels of productivity without inflation or labor unrest? Can it achieve political stability without mass apathy and alienation? Can it create a stable pattern of succession without recourse either to the democratic distemper of the West or the authoritarian repression of the East?

Here the evidence for the success of the norm of illegitimacy is impressive, at least in the arena of economics.

Nations such as Argentina, subject to intense political pressures and a nearly endless round of mass terrorism have nevertheless managed to achieve significant goals. Argentine monetary reserves in 1977 were more than $2 billion, one of the highest in its history. Fair trade, which has been subject to a constant deficit for years, has been improved to the point where a $1.2 billion surplus was achieved in the Argentine balance of payments at the close of 1976. By the close of the last decade, the rate of unemployment in Argentina was reduced to 4 percent. The military regime helped to resolve the nearly chronic state of underemployment as well as unemployment of the Argentine past. And the economic growth rate from 1976 through 1980 has leveled off at 5 percent. Again, whether by accident or design, military rule has served as a stabilizing agency, inflationary spirals notwithstanding.

In Chile the inherited economic problems were more chronic. The inflationary rates have been reduced from 400 percent per annum to less than 15 percent per annum by 1980. By the end of the 1970s, unemployment had been reduced to roughly 5 percent of the work force from an industry-wide norm of 20 percent at the start of the decade. Industrial absenteeism has been reduced from 40 percent to roughly 5 percent in the same decade. In Chile, there is no economic miracle as there has been in Brazil, nor even the stability achieved in Argentina. But despite all sorts of international pressures, the Chilean military junta did achieve a level of economic stability of some consequence.

Calls for human rights have not only largely fallen on deaf ears but have resulted in a summary rejection of American foreign policy pleas, and a direct assertiveness characteristic of successful rather than unsuccessful regimes. This is especially true in Brazil which, by virtue of its economic strength (its rate of increase of gross national product averages more than 10 percent per annum) is able to go directly into the financial market and bypass foreign aid from the United States government. The Brazilian economy is no longer dependent on foreign-aid handouts, and the best possible test of this is that the United States, which in the 1960s provided 35 to 40 percent of all armaments sold to Brazil, and financed additional purchases through dollar grants, today finances only 14 percent of such arms purchases (even though the amounts involved are close to $2 billion). In short, the very successes of military rule and illegitimate political norms have made possible a level of Third World assertiveness inconceivable a decade ago. The norm of illegitimacy, far from signifying the existence of widespread dependency on the United States, on the contrary, has come to illustrate a style of rule intended to create autonomy and independence from the United States.

The same problems exist now as a decade ago, but there is a greater sense of urgency today. Whether or not short-run tendencies can produce long-range economic benefits to large numbers, whether the costs of maintaining a norm of illegitimacy (high military and police power) inhibit economic distribution and fail to expand services, and, finally, whether success at the economic level may engender new classes and social formations, and hence new demands for legitimacy and the institutionalization of authority based on democratic or populist premises—these answers remain as elusive as before. But for the present the norm of illegitimacy has institutionalized military authority and stabilized the mixed economy.

One must conclude in a spirit of caution: certainly in a world in which the United States is viewed as both the *bête noire* of autonomous Latin American development, and at the same time the leading proponent of hemispheric solidarity, the sources of legitimacy and stability are viewed as external. Latin American elites can be expected to demand more in the way of economic development and political nationalism, and less in the way of human rights. This bizarre outcome indicates that we are confronting not only a crisis of legitimation but a crisis in the terminology used to describe the world. As actual priorities are stated, new sources of stress become apparent preventing the exercise of democratic rule.

Elite structures are responsible for both the so-called crisis in legitimacy and the consequent norm of illegitimacy. They often promote ideology at the expense of polity, sectional chaos at the cost of national unity. The central lesson, the basic outcome, of the norm of illegitimacy is, quite possibly, the end of politics, or its relatively complete displacement by policy-making elites. Whether this decline in democratic politics in the Third World is limited to that large chunk of the universe or endemic to the current stage of universal political history is itself an issue to observe and examine—more in trembling than in tranquillity.

10

Military Origins of Third World Dictatorship and Democracy

An unusual consensus seems to be forming, from both Left and Right sources, that in the Third World a swing to fascism is taking place. Different scholars locate different sources of such fascism. For someone like Walter Laqueur such an embryonic movement is endemic to the authoritarian rulerships currently extant;[1] whereas for others like Noam Chomsky the source of such fascism is in the dependence of so much of the Third World upon imperialist powers.[2] The purpose of this chapter is to dispute the assumption itself, whatever its ideological source, and assert that, to the contrary, the motion of these new nations, once their military origins are fully understood, is toward a broad-based democratization along the lines that C. B. Macpherson enunciated: the increase of intra-party participation; the breakup of social rank and class privilege; and increased mobilization and participation of the citizenry.[3] Admittedly, this definitional framework is not perfect, and, indeed, serious questions can be raised about it concerning the distributive concept of democracy. Still, this framework does make possible a discussion of Third World militarism without leaping to the most extreme and mannered conclusions—conclusions that foreclose prematurely and needlessly on the democratic option.

Democracy is an important political issue in the late twentieth century. In part this is so because every other political concept has experienced intellectual exhaustion (or the exhaustion of intellectuals) and in part because the issue of democracy, whatever we mean by the term, has been linked to human-rights concerns in general. Democracy has a compelling force; the drive toward democracy characterizes the entire sweep of the twentieth century. Every major political movement, if it has survived, has evolved toward egalitarianism. Women's movements, racial movements, youth movements, age movements, all are raising questions about egalitarianism. At the same time, questions of liberty have become a central issue. The lynchpin of all these issues is not whether these groups are committed to communism, capitalism, or social welfare, but how each addresses the question of egalitarian and libertarian persuasions.

For many years those concerned with social and political development suffered from an artificial reification: assuming that a society had to surrender democracy in order to develop. Developmentalists were bewitched by the Stalinist notion that to have growth a society had to yield personal liberty. The Third World began to rebel against an inherited European ideology that stated that in order to develop a country must perforce accept the worst dictatorial excesses.

If we labor under the Stalinist notion that development occurs only at the price of democracy, that a society cannot have growth without the sacrifice of large numbers of people to the demons of development, then all is lost. The intellectuals will have lost the battle for the masses on whose behalf they presume to be speaking.

At the outset of the postcolonial epoch, the First World and the Second World were both suspicious of the Third World ideologically as well as functionally; culturally as well as economically. The modernization model—Rostow in economics,[4] Lipset in sociology,[5] Lerner in political theory,[6] Schramm in communications[7]—denied the possibility of "thirdness" per se. The assumption was that the automobile or television was the new absolute measuring-rod of our time. Production measured not only by material goods but by the rapidity of cultural transformation, shifts of knowledge, transfer of technology, and speed of transportation and communication became the touchstone of development. Indeed, the modernization argument went, unless the Third World recognized these changes they would never be on the high road to modernization.

The Soviet response to the "thirdness" of the Third World was to proclaim that these newly emerging nations were really on the historical road to socialism. The USSR supports liberation movements and anti-colonialist movements because they are preconditions to this pre-determined outcome. The Soviets attempted to educate Third World people to understand that their national liberation phase was a *rite de passage* toward the growth of a socialist republic. The rather tawdry model became East Europe. These countries presumably trumpeted their transformation from people's democratic movements to socialist societies. The Soviet/East European approach ultimately denied the notion of "thirdness"—if necessary by force of arms.

If the Third World has a uniqueness, an authenticity, a validity unto itself, conceptually and theoretically, then the question becomes, How do the West and the East cope with its special reality? Can the USA move beyond the modernization thesis; can the USSR get beyond the socialization thesis? The assumption that there are only two worldviews, two world systems, two world powers, and that a country must move from one to the other, has become a dangerous illusion. The United States needs the Third World as an ally and as a recognizable entity, and the Soviet Union and its socialist satellites must also strive to have that world as an ally.

The distinct disadvantage historically and ideologically of Western social and political theory is that it must overcome a history of chauvinism, racism, and colonialism. Yet, it is more feasible for the First World to make that assessment than it is for the Second World. This is so because the Second World cannot easily accept the idea of a Third Way, without the destruction or dismemberment of its fundamental commitment to Marxian theory. The West, by virtue of not having such a solidified theory, by having strategic and policy options that make possible harmonious relationships between the First World and the Third World, can be more flexible as sources of wealth shift. The cataclysmic idea of the "decline of the West" presupposes a zero-sum game that the Third World has increasingly moved away from.

The nature of this interaction between First and Third Worlds, in relation to any given nation or region, is difficult to describe. The role of the United States could become more modest and less bellicose, as the impact of Western technology expands and as mixed economies in relatively open societies grow. Development in major Third World countries like Nigeria, Brazil, and China moves along an axis that grants the possibility of democracy and development. Such a pattern of democratization, even if incomplete in form, would spare the Third World an inevitable nightmare of either Hitlerism or Stalinism.

A critical task of the Third World is to show how the military origins of its systems can still yield democratic form. To be sure, this remarkable transition from uninviting military beginnings to as yet elemental democratic forms has become the central Third World task in the final decades of the twentieth century. This combination of process and policy is one that the United States and the West can only reinforce, not determine or deter. The increase of Soviet aggression in portions of the Middle East and Africa is a response to the fact that the Third World has in fact moved beyond either communism or capitalism into a notion of democratic forms and modes that permit development to take place, and at the same time maintain traditional values.

Development without democracy still dominates Third World processes. For example, despite the intonations of Chile's enemies, the reduction of inflation, elimination of worker absenteeism, and sharp rise in industrial output are impressive. There are limits to what the military can do vis-à-vis social classes. Certainly redistribution of power or property has not occurred in Chile; concentration of wealth has occurred. Chile is an example of a type of free-market Stalinism, characterized by repression, inflation, and reduced consumer spending: the military functions as a surrogate class—a repressive caste rather than a true class.

In Brazil there has been a succession of military rules since the 1964 revolution. The direction has not been toward democratization, but there has evolved a realization of the finite limits of military rule. All ministerial portfolios except the presidency are now civilian in character. The economic ministry operates under the most stringent, balanced economic norms, present inflationary spirals notwithstanding. Brazilian concerns have shifted from raw growth to income redistribution. Present trends indicate a tremendous shift toward a social welfare emphasis, and toward redistribution of resources. Redistribution certainly is not the same as democracy; but it does presume social equity as a goal.

Argentina has been an extraordinary case since the turn of the century. It is a rare early example of economic devolution rather than evolution. At the present time Argentina has entered a protracted process in which economic and political processes are being normalized, but not without great difficulties. The problem, however, is one of legitimacy: Argentina's military has denied in practice what it asserts in principle: the need for civilian government in a free market economy. Having achieved such a high degree of control, institutionalized throughout the century, it is reticent about, some might say incapable of, yielding power. The Argentine military, unlike most other militaries, has developed a powerful network of administrative, bureaucratic, and economic substructures. As a result, its capacity to yield power is limited by the wider interests to which it has become beholden. One might ask if this is characteristic of military regimes, which inhibit the democratization process. The answer, in part at least, involves an appreciation of Argentine political history, a history which witnesses a swing not simply away from militarism but toward constitutional dictatorship.[8] This pendulumlike motion from military to civilian forms of repressive regimes is precisely what may enable Argentina to function as an exception to the general democratization process of the Third World.

While it is difficult to discuss outcomes of those nations newly liberated from colonial conditions, the situation is less problematic with respect to such older nations as Mexico and Turkey. Each in their own way have been prototypical examples of the long term trend toward secularization and civilianization. Just as the twentieth century witnessed a compelling division of labor, so too has it displayed similar sorts of specialization in the political realm. Complex societies display a similar need for managerialism as a bureaucratic style, and this style comes fully equipped with its own forms of rules and regulations—precisely what civilian administrative bodies do better than military leaders. As consensus displaced command in the running of societies such as Mexico and Turkey, the old style military rule, however motivated they might have been by developmental considerations, slowly, often grudgingly, yield ground to civilian forms of rule and quasi democratic norms of political relations. This has been the pattern from Peru in Latin America, Nigeria in Africa, and Maylasia in Asia. This transition from military to civilian structures, while by no means uniform in rapidity or even direction, represents an evolutionary process of considerable significance, impacting international relations no less than the internal organization of Third World nation-states.

It might well be that the capacity of a military regime to function as a handmaiden to the democratization process rests on the balance of economic and political forces in a country at the time of independence. The transition from militarization to civilianization is undoubtedly made simpler by the relatively backward state of class forces in conjunction with the concentration of party forces. In that sense, trends in Africa and the Middle East are probably more conducive to this democratization process than they are in places in Latin America, where political independence took place early, although economic dependence continued over long stretches of time. A good example of the African path to democracy is Nigeria, which, after its civil war in the 1960s over Biafran independence and Ibo separatism, has emerged as perhaps the most important country in black Africa, and where a powerful military rule has made possible normalization of the economy. Party processes and elections have become regularized and the fissures and separations of earlier periods are in the process of being healed.[9]

The costs of military rule are for the first time being seriously compared with the benefits of civilian rule. This re-estimation will not be uniform or smooth. Civilian rule does not necessarily mean democratic decision making. Civilianization is a political process in which the decision-making apparatus is normalized, and in which a linkage to democratic forces is at least possible. Democratization cuts across many ideological barriers; it includes countries as diverse as China, Nigeria, Korea, and Brazil. The process is necessarily going to be different in each country, involving electoral reform in some countries and mass uprisings in others. It is a phenomenon that allows us to characterize the Third World as a structural entity with a unique identity. The Third World does not exist in hothouse isolation from other areas; but it does nonetheless have its own agenda as well as its own evolution. Both the First and Second World approaches are characterized by a tension between the ideology of modernization on the one hand and the ideology of forced industrialization on the other. These approaches assume a teleology, a sense of purpose, alien to the evolutionary processes as such. The Third World is a third structure. The term "non-alignment" conveys a false impression of a Third World located somewhere in limbo, but this is inexact. The Third World is not on the road to anything; it *is* a road. Therefore, the relationships between the three worlds, their quite real contradictions, stem from their relative autonomy. These connections are as important as the relationship of nations to each other.

It is a curious truth that, although many early revolutionary stages in Third World liberation were organized by military personnel, they took quite for granted the fact that in subsequent stages civilian authority, or at least a broader base of political power, would prevail. The penetration of Western technical and intellectual innovations into the systems of work and thought of the armed forces had the effect of broadening their scope and vision. Hence, Kemal Atatürk could speak of the military as "guardians of Turkey's ideals" and not simply a praetorian guard.[10] Likewise, Gamal Abdel Nasser, in speaking of the Egyptian Revolution of 1952, saw in the military intervention a mechanism by which the people of Egypt would "think of governing themselves and having the final word on their destiny."[11] The movement toward civilianization, if not necessarily democratization, in both of these pivotal countries is indicative of how seriously the military took as their mission the transformation of the national culture and not just the passing of political authority.

The armed forces, performing, as they often do, as representatives of the national interest as a whole, have a unique advantage over interest groupings: they are not compelled to perform as representatives of Left, Right, or Center. As such, they are not perceived as part of factional disputes. In those cases where the military itself is aligned with well-articulated trends in the political process, such lofty remoteness is not possible; but in such situations its historic role in the Third World as arbiter of the political and bureaucratic forces within each nation is either obviated or superfluous. The precise tactics to be employed become a source of military frustration. Moshe Lissak, in summarizing the situation in Burma, indicated that military failure stemmed from the over-politicization of the society at large, whereas in Thailand tendencies to fail in its nationalizing mission stemmed from a strictly nonpolitical model of power.[12] Negotiating the orders of power is thus the main task of a Third World military. The problem is that even in those instances, as in the Middle East, where the military is recognized as a focus of solidarity or the embodiment of sacred symbols of society, its high status does not necessarily carry over into the political capacity to negotiate, bargain, broker, or appease a wide array of social issues, i.e. problems of racial and ethnic differentiation, or class and economic advantages.[13] These notorious historical weaknesses are at least as important in propelling civilianizing trends in the Third World as the more abstract, and vague, ideological dispositions spoken of earlier.

The bourgeoisie in Third World countries realize that the military possesses a certain fear of multinationalism. Although marginal segments of the national bourgeoisie are linked with the Left, key sectors of national bourgeoisie power have been internationalized. Economic determinism, however, leads us to a blind spot. That an economic class has a relationship with a military regime does not signify that the economy is the base and the military is the superstructure. Throughout the sophisticated sectors of the Third World the military informed the middle classes of their responsibilities if they expected to continue to have rights. Paying taxes, investing nationally, and sharing equitably created a new sociopolitical reality, a result of military guidance through uncharted political waters.

This emergence of a bureaucratic-authoritarian state does not discount the economic factor in Third World development, but asserts that this state rests upon an uneasy equilibrium between military and political economic forces. It would be a dangerous falsification to think that the Brazilian military coup of 1964 was made in the name of class restoration. The bourgeoisie in Brazil under the military lost that possibility; the popular classes never had the capacity to achieve such an end. Social classes, above and below, make arrangements with the military in order to survive as a class, not to seize power. The ratio of power between the military and the economy is different in the Third World from that in the mature industrialized societies. This ratio gives character and body to the Third World in contrast with the American or European model of economic base and political superstructure. In the Third World the military has an autonomous role. Its power is linked directly to science and industry on the technical side and to the swollen federal bureaucracy on the civilian side.

The military origins of the Third World do not determine the limits of its democratic prospects. This should come as no surprise; the aristocratic origin of the federalists did not decide or determine the limits of American democracy. The military origins of the Brazilian, Nigerian, or Egyptian revolutions no more define the limits of these nations' polity than did the military origins of the Mexican, Cuban, or Chinese revolutions in an earlier age. Militarism only defines a necessary phase that allows for ensuing struggles between democratization and authoritarianism within the Third World. To deny the realities of military power is to dismiss the fact that the Third World has its own integrity—one that cannot be reduced to a West European nineteenth-century model.

Wright Mills's belief that there are three large-scale factors that have primary importance—the political, the economic, and the military—and that these cannot a priori be reduced to each other, is a central element in any serious theory of international stratification.[14] The location of these factors determines their mix. What does this Millsian paradigm look like if we take an international view of political, economic, and military factors, and try to look at the history of the First World, the Second World, and the Third World in terms of the way in which political, economic, and military factors assert their priority at different times?

The First World is characterized by an economic system called capitalism, which came into being long before its political forms took shape in the nineteenth century. The structure of mass politics emerged in modern England in terms of the Reform Acts of 1832 and 1867. These measures gave political shape to the economic system of capitalism by extending the franchise. But the capitalist economy had evolved and was institutionalized between 1640 and 1688. Even the French Revolution assumed a democratic political form at the end of the eighteenth century, although the break-up of the landed gentry in France took place in the early years of that century. The Russian Revolution—the socialist revolution—reversed this causal process. The USSR announced a new economy with a political upheaval. The political vanguard party came into being prior to the economic system called socialism. There was no long precedent to the foundation of the communist bloc. The consolidation of the communist system was a precondition to initiating plans to reach socialist objectives and goals.

The causal relationship between the economy and the polity was largely reversed in the Second World. The Third World transformed the relationship in a different way. Theories have been constructed that abandon the class model itself. In Cuba the theory was called the *foco*; change was said to be based on a focus or locus of power, on centrality of power in groups—groups that presume all kinds of networks that have little to do with a mass or class base. The Chinese Revolution was likewise a model military revolution. In Cuba and China the military was drawn from new strata, unlike the old military of Argentina and Brazil. But new militaries acted like old militaries after they assumed power. A third model, the Nigerian civil war between rival military camps, was resolved by one military faction's victory over another military faction. We should recognize the unique role that the military has played in the Third World and that this factor defines the character of state power there. The *relationship* of political, economic, and military formations defines the character of state power. We cannot determine a priori whether economic power is more important that political or military power. We can, however, chart the expansion of the last at the expense of the former two.

The epoch when the Third World military is propped up by external forces has passed in fact, if not in theory. The military, when it is propped up, is propped up internally by segments of older classes fearing disenfranchisement from multinationals. Even small powers resist United States encroachments. United States denial of military aid because of the repressive, dictatorial, reactionary nature of select Third World regimes has been met with stiff, usually effective, resistance. Private sector support is usually adequate to purchase military hardware on the increasingly active world arms market.

The critical factor is not the source of aid but the character of the military as a class. What are its strengths? What are its weaknesses? From whence does it derive its power? These are extremely complex questions. At one level the military caste is organically rooted to the portion of the industrial-technological complex that creates the means of destruction. The military class is unique in its relationship to the technological structure. The technological order is the command sector that tells the scientific order how to proceed. The relationship of the military to the technological order is the decisive pivot at the level of advanced weaponry. At the level of social formation and social forces, the military derives from a plurality of classes. The specific character of the military sector as such is shaped by World War III industries. The military is in these businesses; but here we should avoid excessive reductionism. While it is true that the military is linked to business, these are often-times failing businesses like Lockheed and Chrysler. That is to say, military systems are characterized not necessarily by their entrepreneurial skills but often by their indifference to entrepreneurial preparedness. NASA has a budget of $6 billion annually, in part because private enterprise has failed to cope with the advanced level of investment such space technology requires. Military classes function best when the state monopolizes industrial infrastructures. Otto Nathan[15] and John Kenneth Galbraith[16] have each offered a prototype of how the modern business economy is determined by the political process. The military is not necessarily interested in profitability—or even in entrepreneurship. The weaknesses of these particular kinds of military-oriented and technology-oriented "businesses" should not be overlooked. On the other hand, military regimes have the capacity to rationalize national economies in ways not easily open to democratic-civilian governments.

The class composition of this military caste does not cut a wide swathe, or have a deep base. Because it is an elite, because it has a narrow base, it cannot sustain legitimacy and is devoid of a system of legitimacy. The imposition of force necessitates the rotation of power, thus creating illegitimacy. The transition to legitimate regimes in the Third World can be institutionalized only by broad masses and broad participation, either by mass action or voter participation. Because the military's base is narrow, it often accepts limits to its own authority, and assumes a need to share in the conduct of power and the transformation of industry.

The relative strength of the military vis-à-vis the political or economic sectors of society is an empirical question, not a matter of including the military as a part of the social formation. One cannot legislate a priori what variables are decisive. The military has the guns; social formations do not. With the exception of the police, which is a unique armed segment within the Third World, the military has a monopoly on weaponry, and hence a monopoly on state power. This is not always true to the same extent; what might be true in Guyana may not be true in its next-door neighbor, Venezuela. But we cannot simply say that any given variable is fundamental and any other is derivative. The alpha and omega of any serious analysis of the democratic process is that the form of democracy does not have to be based on the Westminster model. The civil origins of a society no more contribute to militarism (i.e. fascist systems) than the military origins of a society condemn such a society to dictatorship (i.e. African one-party systems).

There is a rationality to the world military that is similar to the rationality of world economy that some speak about. One need not deny the world historic character of economies, or the world historic character of the polity, in order to assert the world historic character of the military. It is a mistake to think that because social systems differ, as in, say, Cuba or Paraguay, when they have the red bandana flying it necessarily means different things. We don't know how different the Communist Party is from the Colorado Party. That is an empirical question. The differences are no more sharply etched than the similarities. As long as that is true we must look at the military as a unitary phenomenon acting for itself, not simply as a national phenomenon, or a transitional phenomenon, acting on behalf of others. It is dangerous to think in terms of heroic types, or models that dissolve at the point of discovery. We have to look beyond abstract models, to the flag, to the uniform, to the insignia, to the meaning of these symbols. Only by doing so can a sense of the military presence in Third World affairs be understood as the central dynamic it has become.

Although the military is important to the First World and Second World, it is the bourgeoisie, the middle class, that defines the First World, and the political sector, i.e. the Communist Party, the apparatus and the bureaucracy, that defines the Second World. The role of the military has increased in both of those worlds, especially in sophisticated technological hardware. It might very well be that military determinism will not occur in older societies; but the role of the armed forces has certainly increased. What is unique about the Third World military is not merely the hardware component but the social origins of the military and how these regimes come into existence. Historically, the difference is in the character of the social system, defined and determined by the state. The military classes are as crucial to the Third World as the economic classes are to the First World and the political classes to the Second World. However, the class origins or composition of the military condition but do not uniquely determine the democratic prospects of developing areas.

Juan Linz best expresses the larger political context in which military regimes operate. After pointing out that politics is a question not simply of policies and administration, but of appealing to politically interested segments of a society, Linz notes that "large segments of society, still believe, rightly or wrongly, in the desirability of an open, competitive, democratic political system or in the desirability of an ideologically driven, possibly totalitarian society whose elites provide some sense of historical mission to the nation, and thereby satisfy some of the more politically involved citizens."[17] In this setting an authoritarian regime has serious weaknesses. Ultimately, all authoritarian regimes face this legitimacy pull toward the polyarchical model, with political freedom for relatively full participation, or toward the committed, ideological single-party model. The lesson is clear: while one cannot predetermine dictatorial or democratic outcomes, neither should one prejudge the possibility of outcomes simply on the basis of the military origins of many Third World revolutions.

III

Mobilization in the Development Process

11

Capitalism, Communism, and Multinationalism

The overriding ideological posture of the twentieth century has been the contest between capitalism and communism. From it flows the utopian fantasy of one of the chief protagonists: that the contest will be resolved by the triumph of communism and equality over capitalism and inequality. A counter-fantasy excoriates communism for creating a totalitarian nightmare that can be halted only by total allegiance to democratic life as expressed by the "West," a banal euphemism for the capitalist nations.

The rise of a Third World in Africa, Asia, and Latin America, with an attendant plurality of economic forms, political systems, and social doctrines, forced the protagonists to re-examine their fantasies. The rise of multinational corporations primarily loyal to their own industrial growth and financial profitability, rather than to the nation of their origin, further undermined the ideological certainty embodied in a vision of East-West struggle. Multinationalism involves commitments that have taken us beyond premises of a showdown between old capitalism and new communism; or to use the oracular vernacular, classical democracy and modern totalitarianism.

The phase of Third World nation-building that extended from 1945 to 1970 has resulted in a fully crystallized world system. The function and role of the multinational corporations in this world system, and their impact on capitalist-communist relations, are questions of more recent and hence more speculative vintage. Despite the amount of information available on particular multinationals, little attention has been given to how this industrial phenomenon as a power unto itself affects the United States' relations with the Soviet Union, and by extension the structure of capitalism and communism as competing world empires.

The fusion-fission between the United States and the Soviet Union in terms of an evolving multinational system is of direct relevance to the Third World. The penetration of multinational structures and agencies has deeply affected the Third World during the past decade; it has provided new options and complicated old liabilities. The special way in which multinationals rationalize the international economy, especially within the industrialized sectors, is thus of direct and compelling concern to the Third World nations which must maximize their leveraging and bargaining capabilities by the steady review and re-examination of big power relations, especially those between the leaders of the First and Second Worlds.

The Nature of Multinationalism

While definitions of the term multinationalism vary, there is general agreement on the following operational guidelines:[1]

1. Multinationals are corporations that operate in at least six foreign nations.

2. Multinationals are corporations whose foreign subsidiaries account for at least 10 percent of their total assets, sales, or labor force.

3. Multinationals have annual sales or incomes of more than $100 million (which effectively reduces the numer of firms we are dealing with to approximately 200, of which 75 percent are primarily affiliated with the United States).

4. Multinationals have an above-average profit margin and rate of growth when measured against exclusively national firms.

5. Multinationals are most often found in high technology industries, especially those that devote a high proportion of their resources to research, advertising, and marketing.

Economic determinists see nothing particularly novel about multinationals; they see nothing qualitatively different. In short, they view multinationalism simply as imperialism by another name. The difficulty with this position is that it omits everything unique about the present situation.

It is true that the multinational is an old phenomenon that has achieved new dimensions in recent years. Firms like Singer Sewing Machines, National Cash Register, Unilever, and General Motors have been conducting business overseas for many years. The fusion of these older firms with high-technology industrial firms, such as Xerox Corporation, International Business Machines, British Petroleum, Philips, and International Nickel tipped the balance within them from national to international corporate participation. Since each of these industry giants has annual sales that exceed the gross national product of all except several dozen individual nations, the political and economic power they wield is obviously great, although highly diffuse.

What is new about multinationals? In the past these firms maintained classical imperial relations, importing raw materials and exporting finished commodity goods at super-profits. Now they have new arrangements, sharing research and development findings and also patent rights distribution. They manufacture in the economic periphery at lower cost rather than produce the same goods in the cosmopolitan center, which has the additional benefit of quieting nationalist opposition. They develop profit-sharing arrangements between local firms and foreign firms, which involve training and tooling. In some industries a reverse multinationalism is at work, based on raw materials rather than finished goods. Thus, the oil-rich countries of the Arab Middle East form a bargaining collective to do business directly with major oil companies of the West. The result is bartering and bargaining between national governments, such as the Arab oil states joined by Venezuela, Iran, Nigeria, Indonesia, and other members of OPEC, and private-sector multinationals like the powerful oil corporations of America and Western Europe.[2]

What is new about the multinational is not simply the transcendence of the nation-state boundaries to do business, an old ploy of corporations in wealthy nations. More profoundly, it is that they are willing to accept reduced profits through payments of increasingly higher prices for raw materials (like petroleum) and lower prices—hence lower profits—for finished manufactured goods (such as automobiles). This new aspect of multinationals points to the need to modify classical and new forms of Marxism-Leninism alike, since the very essence of politics as a reflexive form of national and economic exploitation is exactly what has been reversed. What we now see is economics as a reflexive form of political exploitation and combination. The relationship of the international economy to the national policy is of considerable interest to the study of multinationalism; however, it is tangential to the purposes of this chapter.[3]

The Buying of Western Capitalism by Eastern Socialism

Post-World War II nationalism prevented any undue optimism about socialism's ability to triumph as a world system and as an international ideology. So intense did nationalist sentiments become in Third World areas of Asia, Africa, and Latin America that the Soviets, after much hesitation, readjusted their policy and ideology and finally recognized a third way, an entity with roots in, but apart from, American and Soviet models.[4] In general terms, the Third World is a cross between a Keynesian economic mechanism and a Leninist political machinery. The dream of an international proletarian revolution, with or without a Soviet vanguard, gave way to more parochial dreams of peoples' democracies and socialist republics that would no more dare try to transcend nationalist sentiments than would the older capitalist regimes in Western Europe. Between 1945 and 1970, the nationalist thrust in the Third World profoundly diminished a belief in the inevitability of a socialist utopia.

When internationalism finally did emerge, it did so in a corporate rather than proletarian guise. The multinational corporation pointed to an international brotherhood of the bourgeoisie and the bureaucracy, to a transcendent class loyalty considerably beyond the national aspirations of even the United States or any other principal capitalist social system. The multinational corporation discredited the socialist utopia as well as the earlier Third World nationalist phase. The multinational is a giant step toward an international economy no less remote from the Second World socialist brotherhood than were national socialisms. The multinationals offered a basket of commodity goods that the Second World desired as much as did the Third World. The relative ease with which multinationals of the capitalist sector penetrated the societies and economies of the socialist sector stands in marked contrast with the difficulties involved in the East and West's concluding the most elemental treaty arrangements at the policy and political level. Doing things in a businesslike way has become a touchstone for rational efficiency as much in the Soviet Union as in the United States. The culture of multinationalism reached Eastern Europe and the Soviet Union long before the actual economic penetration by multinationals; mass consumer demands began to be made by the Soviet public as early as the immediate post-Stalin period.

Extensive trade agreements reached between the United States and the Soviet Union beginning in the 1970s are a gauge of the velocity and the extent of multinational penetration. Can the Soviet Union maintain its basic commitment to production development rather than satisfy consumer demands in the face of foreign business penetration? Obviously the Soviets and now the Chinese think the answer is yes. A consumer orientation means satisfying the immediate needs of a social sector able to pay, and it also means that long-range goals of economic equality at home and fulfilling ambitious goals of national liberation abroad are thought to be less important.

The political potential of multinationals is revealed by their operations in East European nations like Rumania and Hungary. One finds Pepsi Cola Corporation, Hertz Rent-a-Car Agencies, Pan American, and ITT-supported hotels in the center of Bucharest, and, of course, the most conspicuous multinationals in such a city are Western-dominated commercial airlines. Joint marketing ventures are becoming increasingly popular.

> Production may take place at an Eastern plant, a Western plant or both concurrently. As a rule, the communist partner is allotted sales exclusivity in its own and in other Eastern markets, while the Western partner concentrates on the remaining territories. An example is an agreement under which the Simmons Machine Tools Company of Albany, New York, sells in the United States equipment manufactured by Czechoslovakia's Skoda. A new class of contracts calls for one side to assist the other in markets (usually those of less developed countries)

which happens to be more easily accessible to it for geographic, economic or political reasons.[5]

Firms doing business on a licensing basis in East Europe open up channels of communication to the West. The socialist republics' dependence on the Soviet Union is lessened symbolically more than in reality, but it has the effect of displaying the physical presence of the West in Eastern Europe. Beyond that, it permits higher numbers of international conferences at which Westerners participate and interact with participants from China and the Soviet Union. In short, the multinational firm encourages a country like Rumania to strive to become a Switzerland of the East socialist bloc: a place where Israelis, Albanians, Russians, and Chinese meet freely and to the greater benefit of the open-ended socialist regime. In such a context, for one to speak of the corrupting influence of multinationals is to rave like a Puritan divine against autonomy. In the East European setting, national sovereignty is strengthened, not weakened, by the existence of the multinational corporations. This is, of course, at considerable variance with the impact of multinationals in Western Europe. The combined power of Dutch multinationals (Royal Dutch Shell, Unilever, KLM Airlines, Philips) is much stronger in the Netherlands than the government and its various agencies. Indeed, Holland is an example of how national sovereignty can be weakened rather than strengthened by multinationalism.

Raymond Vernon believes that socialism will do as much to promote as to dissuade multinationals from penetrating the socialist and Third World spheres.[6] The idea of a single world market has so deeply permeated Soviet socialism that the USSR must now accept as part of its own economic codebook the rules on any given day of the much reviled free market economy. The Soviets are showing increasing sophistication in the areas of market management and manipulation. In the 1976 wheat deal, the Soviets sent forth twenty buying teams to negotiate independently of one another and without apparent overall coordination. But of course a high level of commercial orchestration enabled the USSR in one fell swoop to fulfill an internal need for wheat, and to buy a surplus amount at low cost for resale on the world market at a high profit. The Soviets in this way have become part of the "paper economy"—for the wheat deal involves the movement of money no less than the transfer of a basic crop.

The Soviets have also done well in the area of natural gas. In 1966, they negotiated with Iran to purchase natural gas that had previously been flared off in the fields because there was no market for it at that time. In turn, the Iranians received Soviet financing and assistance to construct necessary pipelines and associated equipment as well as a steel mill. While the Iranians gained a valuable steel plant, the Soviets began negotiating the sale of the Iranian gas to both East and West European nations. These deals culminated in sales equal to the total Iranian gas supply. The Soviets sold the natural gas for nearly twice the amount they paid Iran for it. Profitable as it was to the Soviets, the deal was equally advantageous for the United States, which avoided paying others more than twice the amount that they had paid the Soviets.[7] Sophisticated multinational dealings across East-West boundaries clearly serve the major powers at the expense of Third World nations.[8]

The Soviet Union and the Eastern bloc COMECON nations (Council for Mutual Economic Assistance) borrow almost exclusively in the Eurodollar market of Europe. American lenders account for a considerable portion of this amount. Hard currency indebtedness of COMECON nations to Western banking interests has soared. By the mid-1970s the totals owed were $22.5 billion. The size of the Soviet economy and its gold reserves make the communist nations a "safe" investment. Soviet and COMECON interest payments buoy Western banking capitalism; while such loans in hard currency buy technology for Eastern communism.

Huge trading blocs regulate relationships between the East and West. The increasing concentration of capital in the West permits the movement of capital from multinational corporations to an interrelated fiscal network that readily connects with the Soviet-interrelated fiscal network.

The Selling of Eastern Socialism to Western Capitalism

The fundamental question within the socialist bloc has been how to develop a modernized society along with a largely successful industrial development. The Soviet Union can mount trips to outer space but cannot satisfy consumer demands for automobiles; it can launch supersonic jet aircraft but cannot supply the accoutrements of personal comfort to make such travel enjoyable. It can mass produce military hardware but cannot manufacture consumer components with a sense of style and design. Every aspect of socialist society betrays the duality between industrialism and consumerism. In this respect the Second World is the opposite of the Third World, where modernization is often purchased at the cost of development; where production is increased with relatively low technology inputs, and produce is exchanged for commodity goods produced in the advanced capitalist sector. Both the Second World and the Third World need and want consumer goods from the First World. The Third World pays for such consumer goods with agrarian goods and mineral wealth, while the Second World wants to pay for such consumer goods with industrial products; it finds a relatively low demand for such finished products elsewhere.

Since the end of the Stalin era in 1952, decreased international tensions have encouraged the development of consumerism in the Soviet Union, which in turn has reinforced tight political, state controls. The assumption is that the stability of socialist bureaucratic regimes will increase as demands for consumer gratification are satisfied. Thus far, this theory has proved correct. Protest has been limited to a narrow stratum of intellectuals who have subsequently been declared malcontents or madmen in the face of general satisfaction with the level of available consumer goods. Whatever the long-term secular trends, and whatever the consequences to socialist legitimacy and class interests, the fact remains that consumer orientation has worked. Multinational penetration of the Second World must therefore be seen as part of a general commitment to achieve political quietude through economic gratification.

The most obvious commodity that the Second World has to sell is technology. The Soviet Union has technological sophistication, built up over more than fifty years of emphasis on industrialization at the expense of nearly every other economic goal. To an increasing degree, American companies looking for ways to reduce their own research and development costs are buying the latest Soviet technology. As the Soviet Union moves closer to a consumer-oriented society its attitudes toward banking and saving have tended to show a corresponding transformation. Since consumer goods are purchased largely through Western-dominated multinationals, the character of international banking communism has drawn closer to that of international banking capitalism, in short, to the essence of banking principles, profits from interests on loans secured by equity arrangements.

The developing First World–Second World banking network has the capacity to rationalize multinational exchanges. Banking capitalism links up with banking socialism precisely because banks are involved in similar international activities and similar investments in profitable enterprises. In the absence of direct industry-to-industry contracts, given East-West structural constraints at the manufacturing level, the banking system pumps life into an East-West economic détente—despite the collapse of political détente in the early 1980s. Arms reduction negotiations can take place even as the political *entente cordiale* of the last decade crumbles.

The rise of a multinational cultural apparatus has been made possible by the widening exchange of scientists, scholars, and artists from East to West and West to East. Such contacts presuppose declining ideological fervor. Both Marxism and Americanism have yielded to considerations of efficiency and effectiveness, and with this yielding has come a vigorous effort to provide methodological guidelines to develop accurate and universally applicable data. New technological developments have potential for simultaneous translation and rapid publication of research results; these, too, bring East and West together. First World–Second World coalescence occurs precisely in areas of intellectual activity relatively uncontaminated by inherited ideological sore points. Such diverse subjects as futurism, computer technology, and machine learning, by virtue of their newness, foster greater joint efforts. Just these areas are most significant from the viewpoint of multinationalist exchanges of goods and services.

Multinationals have made a more significant impact on East-West accommodation than any other influence. Led by the United States and the Soviet Union, scientific academies of a dozen nations have set up a "think-tank" to seek solutions to problems created by industrialization and urbanization. Such problems as pollution control, public health, and overpopulation are to be studied by an International Institute of Applied Systems Analysis with overseas headquarters in Vienna.

Multinationalism and the End of Classical Imperialism

Multinationalism corresponds with a concomitant transformation of imperialist relations. Classical direct military occupation of foreign territories characterized a colonial period in which modernization was low. By the advent of the twentieth century, the export of banking and industrial wealth owned by advanced powers was exchanged for the mineral wealth and natural resources of the peripheral, or colonized area. Economic arguments concerning the relative value of agricultural and industrial products aside, the historical fact is that underdeveloped areas were, and still are, characterized by an agrarian base that is dependent on the export of raw materials and the import of finished goods and commodities. Over time this pattern began to break down, the first piece of evidence being the failure of the masses to participate in the selective distribution of commodities, creating huge riots and revolutions in overseas developing areas. Thus, contradictions between the national middle classes and the rest of these underdeveloped societies subjected classical imperialism to intense pressures. National liberation and socialist movements of various types and structures simply invalidated the classical model of colonialism.

To stop the erosion of their international position, the imperial powers altered their strategies and, beyond that, their profit picture. Marx pointed out in *Das Kapital* that in a domestic context the percentage of profitability is less decisive than maintaining profits. The increase in product use is a means of stabilizing or increasing profits in absolute monetary amounts rather than in percentage measurements. Hence, capital-intensive industries permit less overt exploitation without systemic collapse. The same phenomenon has occurred in the international context: a shift occurs lowering the percent of profits but not the actual flow of funds. Multinational corporations invite local participation in factory and industrial management, train local talent in tasks requiring special technical competence, move toward joint ownership with local middle classes or local bureaucratic classes (if dealings are with a socialist country), and transfer factories and technicians when necessary. In short, they do everything except surrender their positions as profit-making units operating in an overseas climate. This overseas climate has become increasingly antagonistic to the signs and symbols of imperial enterprises, while desperately demanding more of the goods and services of these same foreign firms.[9] Multinationals help bridge the gap between revolutionary nationalism and establishment internationalism in Third World societies by acquiescing to the symbolic demands of nationalists and revolutionists, while at the same time satisfying the very real economic demands of the conservative middle classes.

The multinational corporation has given increased weight to Lenin's initial focus on imperialism, albeit in an unforeseen manner. At the turn of the twentieth century, the imperial powers engaged in banking-industrial capitalism were the United States, England, France, Germany, Japan, and Russia. After the Russian Revolution, the Marxists postulated that Russia would be out of the imperialist orbit for all time. After World War I they further postulated that the imperialist powers would redivide Europe so as to limit and minimize German participation in the Imperial Club. After World War II Marxists further deduced that the back of Western European capitalism had been broken, and that certainly both German and Japanese capitalism had been brought to heel. A quarter-century later, we have witnessed a Grand Restoration, of which the Marshall Plan was the true precursor and advance guard of the present multinational system. The same cluster of nations that prevailed at the turn of the century today control the overwhelming bulk of the international economy. The economic mix has changed; it is now far more favorable to the United States than it was in 1900, and, curiously, much more favorable to Japan and Germany.

The staying power of dominant nations has been quite durable, whatever the rotation of political power at any given timespan within the twentieth century. Perhaps all political revolutions share a similar concern for the maintenance and eventual extension of economic domination. By altering the fundamental relationships between the bourgeoisie of advanced countries and the bourgeoisie of Third World countries, multinationals have changed the terms of the international game. Too much Marxist thinking in the current era supposes that dependence and underdevelopment are the handmaidens of backwardness. In fact increasingly the Third

World exhibits a correlation of dependence *and* development. This is as true of Soviet penetration in Cuba as it is of United States penetration in Brazil; the senior and junior partner arrangement more accurately describes present developmental realities than does the conventional model of superordination and subordination. By internationalizing capital relations, multinationals have also internationalized class relationships. Obviously, the situation with Soviet satellites is more complex, since trade and aid relations are filtered through a grid of political and military tradeoffs; yet the same principle clearly obtains. Multinationalism permits development yet maintains a pattern of benign dependence.

The Political Economy of Multinationalism

Perhaps the least anticipated consequence of the multinational corporation is the reappearance of militant unionism. Worker resistance to the multinational's attempt to seek out cheap labor as well as cheap raw materials is clearly on the rise. Highly paid West German optical workers must compete against low-paid workers from the same industries in Eastern Europe. Auto workers in Western Europe find themselves competing against workers in Latin America who produce essentially the same cars. Chemical plants of wholly owned United States subsidiaries are put up in Belgium and England to capitalize on the cheaper wage scales of European chemical workers and to gain greater proximity to retail markets. The United States increasingly has become an investment bargain: witness the establishment of Volkswagen plants and the purchases by Europeans of established United States companies.

Directors of multinational companies argue that they can take advantage of the protectionist system of closed markets in the United States while pursuing an anti-protectionist approach abroad, thereby deriving the payoffs of selling to the American worker at American market prices while employing workers overseas at lower European wage scales. Investment abroad is also a way to get beyond US anti-trust laws that do not apply in other countries. As Gus Tyler observed: "For all these reasons—cheap labor, tax advantages, protected markets, monopoly control—as well as for other reasons of proximity to materials or markets, the giant conglomerates of America are moving their investments massively overseas. The result has been a rising threat to American employment and trade: jobs have not kept up with either our growing population or market; exports have not kept up with expanding world trade."[10]

Whatever the merits or demerits of the rationalizing capacity of multinationals, a partially revivified working class has been created, with greater class solidarity and cross-class national solidarity than in earlier periods. In the major wars of the twentieth century, the working classes consistently lined up behind nationalism and patriotism and in so doing have frustrated just about overy prediction made on their behalf by left-wing intellectuals. Now, when so much left-oriented rhetoric has become infused with an anti-working-class bias, proletarian militance is emerging as a function of self-interest rather than ideology.

Working-class life is still organized along national lines; but when confronted with middle-class internationalism as represented by the multinationals, it must either create new trade union mechanisms or revitalize old and existing ones. For example, a relatively insular trade union movement such as the British Trades Union Congress (TUC) has vigorously responded to multinationals as a threat. It has demanded union recognition as a precondition for setting up foreign subsidiaries in the United Kingdom; and furthermore it has asked organizations such as the Organization for Economic Cooperation and Development (OECD) to channel working-class demands on multinationals.[11] But while British responses have been legalist and proffered through government agencies, European workers on the Continent have become more direct and forthright in their dealings with multinational-led strike actions and corporate lockouts.

Renewed working-class activity has had a stunning effect on East-West trade union relations. It is axiomatic that socialism does not tolerate or permit strikes since, in the doctrine of its founders, socialism is a workers' society, and a strike against the government is a strike against one's own interest. That such reasoning is a palpable hoax has never been denied. Indeed, the leaders of Czechoslovakia, Hungary, and other East European states have become quite sensitive to such mass pressure from below. Yet strike actions in the East have been rare, and have met most often with repressive measures. Only in the 1980s, and with dramatic consequences, has the Polish working class asserted its class role over and against the party role. Working-class international action between laborers in "capitalist" and "socialist" countries has been virtually non-existent. But these deep political inhibitions are breaking down. Given accelerated organizational efforts along class rather than national lines, we may be entering an era of working-class collaboration across systemic lines, not unlike the coalescence between the bourgeois West and bureaucratic East.[12]

Developing proletarian internationalism, for example, the support rendered Poland's solidarity by the AFL-CIO in the United States, cuts across national lines for one of the few times in the twentieth century. It cuts cross systemic lines, being less responsive now to Cold War calls for free labor or socialist labor than it ever was at any earlier time in the post-World War II period. The vanguard role in this effort is being assumed by the workers in the better-paid and better-organized sectors of labor—in the specialized craft sector more than in the assembly-line industrial sector. While new mechanisms are being created to deal with multinational corporations, the more customary approach is to strengthen the bargaining position of available organizations, such as the International Metal Workers' Federation and the International Federation of Chemical and General Workers' Unions.

What we have then is an intensification of class friction but on a scale and magnitude quite beyond the conventional national constraints of the past. It is still difficult to demonstrate or to predict whether such class struggles can be resolved short of revolution in international industrial contexts, as they were in the national areas. If Marxism as a triumphant march of socialism throughout the world has been thoroughly discredited, it still manages to rise, phoenixlike, out of the bitter ashes of such disrepair through the internationalization of groups dedicated to organized terror. The intensification of class struggle at the international level remains muted by the comparative advantages of multinationalism to countries like Japan and the United States. But if such comparative advantages dissipate themselves over time (and this is beginning to happen as less-developed nations begin to catch up), then class competition might well intensify—both as legitimate mass movements, or, failing that, as illegitimate terrorist movements.

Big-Power Convergence and Multinational Realities

Multinationalism has refocused attention on the theory of convergence—that particular set of assumptions that holds that over time the industrial and urbanizing tendencies of the United States and the Soviet Union will prevail over systemic and ideological differences and will converge or at least show enough similarity to prevent major grave international confrontations.[13] The convergence theory does not postulate that the two systems will become identical but rather that a sort of political twin-track coalition network will take place. Convergence more nearly represents a parallel development than true merging. In this sense, multinational corporation interpenetration is quite distinct from convergence, since the two superstates at the functional economic levels link, but disparity at the political organization level continues. In an interpenetration such as that brought about by multinationals, systems of society do indeed meet and cross over. The lines of intersection are clearly evident as the data show. The implications of such a development are far reaching.

The evidence for the convergence theory has been strengthened by the rise of multinationals. Without entering into an arid debate about whether capitalism or socialism can remain pure and noble, the facts are clear enough. The United States and the Soviet Union have shown a remarkable propensity to fuse their interests at the economic level even as their differences at the diplomatic level are exacerbated. Even if détente has been temporarily shelved, the impulse to agreement remains intact. Indeed, the doctrine of national self-interest has been superseded by one of regional and even hemispheric spheres of domination by the two world superpowers.[14] Friction occurs when these spheres of interest are ignored or violated.

The concept of systemic convergence is certainly not new. The pros and cons of the debate have been articulated by intellectuals and politicians alike. Commonalities between the major political and economic powers have long been evident. Geographical size, racial and religious similarities, even psychological properties of the peoples of the USA and the USSR, all conspire to fuse American and Soviet interests. Are such commonalities sufficient to overcome long-standing differences in the economic organization of society, ideological commitments, and political systems of domination? This question remains largely unanswerable as long as the mechanism for expressing any functional convergence is absent. The unique contribution of multinationalism to the debate about convergence is precisely its functional rationality, its place in contemporary history as the Archimedean lever lifting both nations out of the Cold War. Multinational links take precedence over political differences in prosaic but meaningful ways. They rationalize and standardize international economic relationships. They require perfect interchangeability of parts; a uniform system of weights and measurements; common auditing languages for expression of world trade and commerce; standard codes for aircraft and airports, telephonic and telegraphic communications; and banking rules and regulations that are adhered to by all nations. Convergence takes place not so much by ideological proclamation (although there has even been some of this) but primarily by organizational *fiat*. Ideological differences are held constant while every other factor in international relations is rotated.

Even as we enter the multinational era, questions arise as to the efficacy of this resolution for world society. Some critics believe the social structures of modern business are in contradiction to the larger value complex of society.[15] Some crusaders see the multinational corporation as the beginning of a true internationalism.[16] Some have suggested that multinationals may resolve certain issues in relationship to the underdeveloped regions of the Third World as a whole. Beyond the multinationals, for example, may be regional organizations serviced by public institutions. Regionalism already exists in the form of the European Common Market, the Latin American Free Trade Federation, COMECON, etc. These agencies, while serving the countries who are member-states, reveal their indebtedness to political pressures of power blocs that the multinationals move beyond. Opposition to multinationals thus becomes a special variant of world political readjustment. And if opposition keeps the multinationals in check, it does so at the expense of any real movement beyond the current political immobilism. If multinationalism is not exactly the fulfillment of an egalitarian dream, the return to more parochial forms of socioeconomic organizations is no real improvement; and furthermore, will not be accepted either by recipient or donor nations.

Multinationalism has played a major role in establishing détente as the central thrust in big-power settlements. In doing so, the bargaining power of smaller nations vis-à-vis the superpowers has been lessened. But, whatever problems this leaves in its wake, this myth-breaking development at least makes possible a more realistic international political climate.

This new multinational arrangement among the superpowers is a repudiation of the earlier moral absolutism of anti-communism and anti-capitalism. In a sense, and one step beyond an acknowledged end of the Cold War, this geopolitical economic redistribution may also solve a major problem of the multinational corporation, its transcendence of the limits and encumbrances placed by national sovereignty. By an internationl linkage of the superpowers, the problems of multinational regulation, which loom so large in the established literature, can be tackled, if not entirely resolved, by appeals to commercial reality rather than political sovereignty.

The thesis presented by George Kennan that the end of the Cold War came about as a result of a series of victories of the United States over the Soviet Union is simply untenable.[17] Stalinism was remarkably legalistic in its foreign policy, whatever its internal extra-legalities were. Beyond that, the Soviet Union in its post-Stalinist phase has neither shrunk nor disappeared. Current Soviet policy, especially as it affects Eastern Europe, can only be described as extremely aggressive. Its military power is now second to none. It is precisely the absence of a victor, the existence of a relative stalemate, that prevents the major powers from aggressively pursuing their collision course—a course that could threaten both empires at the expense of outside factions in the Third World (China, and even non-aligned nationals like India)—waiting in the wings to pick up the pieces. The fear of loss rather than the challenge of victory is the cement holding East-West relations intact.

Advances in both the political and social realms have come about in rapid succession as a result of multinational economics: modest arms control agreements, exchanges of research and development technology in basic fields, and even sharing certain costly information in areas of space research. These have signaled the termination of the Cold War, even if it has not produced détente. Multinationalism, in its extra-national capacities, has rationalized this new foreign-policy posture on both sides. Terms like "have" versus "have-not" states have replaced an older rhetoric of capitalism versus socialism. Precisely because the Cold War has not been resolved through victory by either side, the elite leadership in powerful states has felt that the coalition of the big against the small, of wealthy nations against impoverished, and even of white-led nations against non-white-led nations can guarantee the peace of the world and the tranquillity of potential sources of rival power like China in the East or Germany and France in the West. The mutual winding down of the Cold War in favor of cool rhetoric resolves rivalries among the major powers and consolidates the subordinate position of the Third World for the duration of the century. The economic mechanism for this new shift in fundamental policy is the multinational corporation.

Whether this cement will stay firm remains in grave doubt. There are special problems in any Soviet participation in multinational affairs, particularly as they relate to sub-national activities in which market values are directly pegged to supply and demand. The peculiarities of a controlled or planned economic system are such that purchasing habits are more readily manipulated than in open market conditions. However, if this is an advantage in relations between nations, such advantages dissolve at strictly entrepreneurial levels. Currency values in the Second World are uniformly soft, and just about worthless in an open banking market. The worth of its goods is likewise hard to establish, except in comparative contexts; hence if equipment is purchased from the Soviet Union, the value of such goods if often determined by similar goods (or services) available in the First World sector. There is also a debasement of the quality of goods that makes Third World purchases of finished goods from the Second World less valuable in qualitative terms than comparable goods from First World multinationals.

Second World participation in multinational activities, while abstractly offering virtually unlimited market potentials, is in fact quite limited. The best arrangements involve joint stock participation, as in automobile leasing or hotel construction; or the outright purchase or sale of agricultural commodities or minerals where issues of quality rarely arise. But the "dangers" of such unabashed consumerism have long been recognized by Soviet authorities.

Hence, while Second World penetration of the multinational world is genuine, the actual extent of First World penetration of the Second World still remains the foundation of this real, but asymmetrical, relationship between players in the great multinational contest. This is another way of saying that the national factor (the United States versus the Soviet Union) no less than the systemic factor (capitalism versus socialism) remain potent in the determination of both multinational penetration and interpenetration. Beyond this globalization of bilateralism is the awakening and emergence of a Third World that is singularly unresponsive to joining alliances that would minimize its own options.

12

Human Rights and the Developmental Process

Many of the slogans of our times are painted on a Third World canvas. The campaign for human rights concerns violations, more often than not having to do with incarceration and assassination in Third World nations. This in itself creates an element of stress, with human rights being a judgment as well as a policy made by First World countries in respect to Third World countries. In a century scarred by massive destruction in the advanced industrial powers, the slogan of human rights has a prima facie liability attached to it. It is a policy initiated by those least in a position to assert moral superiority. That the anomaly of such human rights initiatives emanating from the First World and imposed upon the Third World has been muted is often due to the weakness of those military regimes that have borne the brunt of this policy initiative. It remains a troublesome element in such policy how claims of human rights violations in stable totalitarian regimes such as Cuba can be subject to the same set of social sanctions as those of an unstable, but conservative, character. The virtual silence of the Soviet Union on this subject since the original Helsinki accords would indicate that the human rights tactic remains a First World response to Second World totalitarianism no less than Third World repression. The universalist plea for the restoration of human rights for all those victimized in whatever part of the world violations occur cannot disguise the strategic particularism involved. From the very outset, the subject of human rights within an international context has been plagued by uncertainty and ambiguity.

The Pivot of Rights and Obligations

The widespread disregard, even disdain, for a concept of social responsibility paralleling that of human rights generated a kind of interest-group politics in the seventies that tended to sacrifice the whole (national interest) for the parts (specific interests), all but destroying a notion of community in favor of regional, local, even entirely egotistical "issues." Single-interest politics and its attendant special-issue lobbying efforts, while serving as effective mobilizing devices, have immeasurably marked the political process. In the United States, political statements on matters ranging from taxation to education to defense policy became instruments by which public officials could be deposed or elevated. A government of laws became one of lawyers; the force of institutional affiliation was eroded and transformed into a baser force of individual influence. To perceive of obligations as anything beyond quid pro quo became absurd in a perfect exchange system.

As human rights became a demand for sameness, a calculus of statistical claims of parity, in which any reward differential within a society becames a reason for public outcry, the question of human rights became one of private avarice. Demands for social and economic parity raised serious questions about whether a national structure could exist, or, more specifically, whether incentive and reward could be recognized. The transvaluation of the human rights issue into one of perfect equality holds open the spectre of an ideology and a system in which democracy secured through totalitarianism is instead absorbed by it, and quickly dispensed with in favor of restorationism.

With human rights, as with other public concerns, political events often dictated public discourse. Since the founding of the United Nations, in the mid-1940s, the question of human rights has remained in the province of UNESCO conferences. Curiously, only in the late 1970s did it emerge as a central issue.[1] This new sense of concern was a result not of an intellectual breakthrough but rather of the development of a concern for rights as a major instrument of American foreign policy. The passion for human rights took on accelerated form only as a response to the weakest point in the Soviet ideological arsernal coupled with the weakest point in the American military arsenal.

Yet one should not be cynical about the subject of human rights simply because the sense of concern was so clearly related to national and international policy processes. The War on Poverty was a political invention long before the emergence of a literature on poverty; and even though that particular rhetorical war has long since passed into oblivion, the realities of poverty remain. The major issues of the century are placed on the agenda of public discourse often through political formulas and slogans. This provides insight into how mobilization is achieved through such phrases as human rights. The conversion of such cheap phraseology into expensive policy initiatives remains part and parcel of the arsenal of interest-nation politics.

Politics is a game of vulnerabilities, and the human rights issue is clearly where the socialist Second World has proved most vulnerable, just as the economic rights issue is where the capitalist First World is most open to criticism. The interplay of forces, the competition of world historic systems and empires, provides an opportunity for ordinary citizens to gain marginal advantage over the systems they inhabit. The debate on human rights can be conceptualized in part as a struggle between eighteenth-century libertarian persuasions and nineteenth-century egalitarian beliefs—that is, from a vision of human rights having to do with the right of individual justice before the law to a recognition of the rights of individuals to social security and equitable conditions of work and standards of living. Whether human rights is essentially a political or economic concern is not a serious issue. It is enough to formulate an accountability system monitoring and evaluating politics and economics.

The right of justice, or the right to a formal education, is easy enough to absorb within the framework of almost any social system. But when rights are carefully defined in terms of costs, when freedom of belief becomes translated into freedom to impart information and ideas without harassment, when social security is translated into old age insurance, when rights to privacy are viewed as the right of every individual to communicate in secrecy, when social rights are translated into the rights of mothers and children to special care and protection, when rights to work involve the right to form and join trade unions and the right to strike, when rights to personal security involve measures to protect the safety of conscientious objectors, when rights to fair trials include protection against arbitrary arrest or detention—then the entire panoply of rights assumes an exact meaning that it otherwise would not have.

The widespread interest in human rights often reflects the absence of these rights in many parts of the world. There is a great deal of concern on matters of cruel, inhuman, or degrading treatment because there is so much cruelty, inhumanity, and degradation present in world affairs. There is concern about the rights of self-determination because there are so many violations of those rights in the name of national integration. International law calls for the punishment of genocide because the twentieth century has seen the alarming development of genocide for statist ends.

The dichotomy between practices and principles, between reality and rhetoric, gave the human rights issue its special volatility. Yet the one enormous breakthrough that has evolved over the century is the sense of right and wrong. A common legacy of both democratic and socialist systems of marketing and planning, of libertarian and egalitarian frameworks, is the assumption that there is such a goal as human rights. Only 150 years ago slavery and serfdom were a vital force in human affairs, and wars were fought to protect chattel slavery as states' rights. The extent and velocity—at least conceptually—of how far we have come are dramatic.

The central characteristic of the twentieth century, which profoundly demarcates it ideologically from previous centuries, is that rights are presumed to be inalienable. Institutions in the past have been largely concerned with theories of human obligation: what individuals and collectivities owe to their societies and to their states—an automatic presumption that the citizen has an obligation to fight in wars whatever the purpose of the war, or the notion that economic failure is a mark of individual shortcoming rather than societal breakdown. The hallmark of the twentieth century is that the question of obligations has been placed upon the shoulders of institutions rather than upon those of individuals.

There are risks in this transvaluation. One might well argue that the tilt has turned into a rout; that issues of the duties of individuals to the community, or the limitations of human rights to ensure national survival, have not received proper attention; that social research has so emphasized the minutiae of imbalances of every sort that even homicides are now blamed on television violence. All transvaluations carry the potential of hyperbole and exaggeration, and there is danger of throwing out the baby with the bathwater. History has been written in terms of dynasties, nations, and empires. As long as that was the case, human rights hardly counted. Only now, when these larger-than-life institutions—these dynasties, nations, and empires—are dissolving, can it be seen that the individual is the centerpiece of all human rights and that the expression of these rights must always remain the province of the free conscience of a free individual. In this very special sense ours is a century in which individualism has emerged beyond the wildest imaginings of previous centuries. Paradoxically, it is also the century of the most barbaric collectivisms which have put into sharp and painful relief the subject of human rights.

Choosing Between Rights and Goals

The most meaningful expression of human rights involves the globalization of the moral economy, that is to say, the idea that the goods and services available to each individual should be maximized and that such pursuits should be followed with a certain freedom from interference, especially by authorities in power. Human rights may or may not be enshrined in a constitutional framework, the bases of such respect for rights may be normative or pragmatic, but they are widely shared and recognized. Indeed, human rights as a concept derives from a strongly twentieth-century idea that every individual should count as one, no more or less than one; furthermore, that profund extremes of wealth and poverty only limit the exercise of such rights by completely imbalancing rights and responsibilities.

Quite beyond the issue of balancing rights and responsibilities—more specifically, human rights and national responsibilities—lie such large-scale philosophical issues, as Sidney Hook properly noted, that frequently translate themselves into the question of whose ox is being gored.[2] Claims for universality of judgment and decision making in employment meet with contextual claims of historic segregation and therefore require affirmative action to bring about rapid adjustment. Trade union arguments for open shops that permit unionization yield to demands for closed shop guarantees. Curiously, management becomes the champion for open shop approaches to unionization. Such examples can be multiplied a hundredfold throughout the world. What is at stake is a realization that rights and interests are not in perfect harmony, and, as a result, demands for a human rights posture inevitably engender counter-thrusts for a self-interest or national interest posture. For this reason, moral claims for a human-rights posture cannot be effectively made without a parallel realization of empirical claims for a national-rights posture. Nevertheless the relativity of rights should itself be viewed as a hallmark of twentieth-century progress.

The most troublesome problem in the human-rights program is a myopic incapacity to appreciate the degree to which some rights conflict with other rights: the right to know versus the right to privacy, the rights of private property versus the rights of each person to be free of slavery, and so on. In short, every right is empirically circumscribed. Laws do not so much mandate rights as they circumscribe such rights in the name of high moral truths. As a result, to speak of a human rights policy is to assume the unitary character of rights, which is precisely what is most seriously open to challenge and, assuredly, to doubt. There is even a certain question as to whether individuals or societies are the ultimate repositories for rights. For it must be recognized that individual and collective goods are not mutually negotiable. All of this is said not to frustrate the search for a human rights policy, but, better, to recognize that the subject of human rights is, to be sure, a matter of policy, and not some set of precepts written on tablets for all times for each person to hold sacred.

Each right has its limit. The right of each nation to self-determination is limited by the right of a sovereign to exercise force to maintain its cohesiveness, such as the justification invoked by the Nigerian central government to suppress the Biafran uprising. The right of each nation to have safe and secure borders is circumscribed by the practice of genocide of its own people: witness the invasion by Vietnam of Cambodia; or more clearly, the invasion of Uganda by Tanzania and Kenya. Again, such illustrations can be multiplied, but the point must be made, even repeated, that human rights are not a monopoly of the First World to be doled out to the Third World; nor can human rights be discussed at such a high level of abstraction that the specifics of rights and responsibilities are buried by consummate rhetoric. Advanced Third World nations must be made a central part of any paradigm concerning human rights—not as recipients of received wisdom, but as donors to any world stability.

Human Rights as a National Policy

Human rights programs, like the New Deal, the Fair Deal, and the Great Society, represent a historical tendency on the part of the American executive government to give distinction and individual character to each of their administrations. The Democratic party in particular has traditionally felt a need to delineate itself in contrast with previous administrations. In part, this is a consequence of a breakdown of parties per se in America and the substitution of political formulas based on mass communication in their stead. The human rights issue as it now stands, in contrast with previous executive political rallying cries, differs in its international dimension. In the past, most major slogans were confined to the national political system. Human rights are located in the international arena. In part, this is deep recognition that there is a high correlation between war and the adoption of social measures intended to ameliorate world poverty and discrimination. Indeed, as Robert Nisbet reminds us, our highly prized social gains have been the result more of the outcomes of war than the ideology of rights.[3]

In the 1950s, the key term was modernization, easily enough defined as a maximization of the production of consumer goods and human resources. In fact, in operational terms modernization was strongly equated with transportation and communication, or those areas in which the United States was a world leader. In the 1960s, with the growing apprehension that material abundance, or modernization, might actually be counter-productive to national goals, because it left intact the uneven distribution and class characteristics of such consumer impulses, the key phrase became egalitarianism—or the drive toward the relatively equal distribution of world resources and goods. But by the end of the 1960s, a new difficulty soon become apparent: as Third World nations were urged to adopt egalitarianism as an international ideology, what had been a virtual First World monopoly of the material conditions of abundance was starting to evaporate. The drive for equality at the domestic level became more intense and problematic. The costs of international egalitarianism were more easily borne than the realization of domestic tranquillity.

There were other domestic problems in driving the equity stick too hard: less innovation, higher taxation, and a huge shift in moral specifications. Above all, egalitarianism came to mean that growth itself had a higher price than many environmentalists and social ecologists were willing to pay. When the growth curve leveled off, and even began moving downward as a result of energy shortfalls and oil boycotts, the slogans of the 1960s became increasingly dubious as benchmarks for policy making in the 1970s.

The turn toward human rights in the seventies was an American policy presentation that found it useful to convert equity demands into liberty demands. It was the strong suit of the West in an era of declining American hegemony and a corresponding growth in Soviet power. The human rights issue, covertly for the most part, celebrated the fact of high political and social freedom in the West—just as modernization in the 1950s celebrated the fact that America was a consumer society, and egalitarianism in the 1960s celebrated the fact that large numbers of people shared in this largesse. Like political formulas of previous decades, the human rights issue provided a sharp contrast to the socialist sector, with its prima facie constraints on human rights: from exacting harsh punishment for small crimes, to refusal to grant travel or immigration visas. Thus, human rights in the West, whether measured by press standards, due process of law, freedom of worship, voluntary association, or multiple parties, offered a contrast to Soviet power and provided an illustration of a certain inability of the slogans of the previous decade to transform themselves into reality.

Any newspaper on any given day makes it evident that human rights, however potent as a normative instrument of foreign policy, are constantly tempered and even temporized by reality. Stories from a single issue of the *Financial Times of London*[4] illustrate this point: the United States, which had urged sanctions against the Soviet Union for its persecution and imprisonment of dissidents and human rights policies generally relented on the sale of drill bit equipment and technology vital for the Soviet oil industry. Such equipment represented a considerable financial windfall for the United States, and it also meant that the Soviet Union was spared from generating new forms of research in a complex field. On the other hand, when the United States confronted a far weaker nation, such as Argentina, it held up the sale of Boeing jetliners to Argentina because of its alleged human rights violations. As a result, on the same day that the United States sold equipment vital to the Soviet oil industry—with a pained admission that this violated human rights considerations—it denied a jet package to Argentina on precisely the same human rights grounds.

When one thinks of the human rights question in the context of international economic reality, it becomes apparent that there are limits to the implementation of any policy, especially by a United States weakened by capitalist competition even more than socialist victories in the late seventies. Vacuums are filled; economic vacuums are filled rapidly. Thus, while the United States denied certain loans to Brazil, again on the basis of presumed human rights violations, the Japanese and Italians were more than willing to fill this gap, making a $700 million arrangement whereby Japanese and Italian capital provided long-term financing for new steel projects in Brazil, in addition to which the three companies—Sidebras, Kawasaki, and Findsider of Italy—provided terms that assisted Brazilian capital needs on more favorable terms than past United States loans. This sort of competitive capitalist behavior became standard international practice.

For many countries human rights is not a particularly important constraint. For example, Pakistan, one of the more overt violators of human rights for its own people, developed close links with the Muslim countries and with the Shah's Iran, with the former Iranian government underwriting highway construction to wheat importation in Pakistan. In that situation, not only was Iran helping Pakistan but the orchestration of this pact was done through the Eurodollars provided by none other than the American Citibank Corporation. The bank also helped set the rate of interest to be charged to Pakistan by Iran. These technicalities are

replicated and later repudiated every day of the year. A
it is impossible to speak of the implementation of a hu
policy as if it were a mathematical axiom. It might well
politics of boycott is the ultimate expression of a policy of human rights. But this presupposes a total monopoly of goods and services, which the First World does not now possess, if it ever did.

It is appropriate that the West appreciates one particularly important concept taken from the arsenal of Soviet policymaking, which involves figuring out the main danger, the primary and secondary forms of constraints and contradictions. The simple moralism that would deny aid to both El Salvador and Brazil because of human rights violations fails to account for the differential importance of the two nations in the maintenance of United States foreign policy; and no less their differential capacity to defend themselves against internal rebellion. As a result, even if there is a basic human rights policy, implementation tends to be differentiated if it is to make any sense, or if it is not simply to be ignored as a political spoof.

Human Rights as a Global Issue

Human rights, at least on a global perspective, has been linked to choices between political deprivation, presumably characteristic of the Second World, and economic exploitation, presumably characteristic of the First World. Aside from the reification and polarization involved, the problem is that the human rights issue is thrown back upon ideological grounds. One senses a growing apathy with human rights questions because that rhetoric has not brought about desired changes; on the contrary, it has sometimes caused a restoration of more harsh and punitive measures in countries feeling threatened.

A fitting and proper role of social science in the human rights field is to move beyond broad abstractions and seek out concrete expressions of both the exercise and abridgment of human rights. In this regard, the human rights issue can be joined to a framework larger than itself and politically more significant. It can be fused to questions about social relations, political regimes, and economic systems.

Instead of talking abstractly about the right to work, there is a need to discuss concretely conditions of work, protection of migrant workers, occupational work and safety measures, and social services for employees. Instead of talking about right to life and liberty, policy recommendations can be made specifically about protection under the law, a right to a fair trial, and the security of a person who is incarcerated. If we are talking about political rights, questions should be raised about conditions of voting, levels of participation, numbers of parties permitted to contend, electoral expenses, and the role of local vis-à-vis national government. Instead of talking about citizen rights, issues can be raised about the conditions of migration outside a nation, freedom of movement within a nation, rights of asylum, protection against deportation, and the character of national and ethnic affiliations.[5]

There are national differences and these cannot easily be eliminated. Through a sophisticated series of social indicators, researchers can provide some muscle and not just more flab to the human rights issue. For example, the Yugoslavs have extended questions of the right to work to include free choice of employment and conditions of employment, protection against unemployment, equal pay for equal work, the right to favorable remuneration, and even the right to form trade unions. On the other hand, the Yugoslav program does not include the right to strike. So at this time, the need of the research community is to achieve a stage beyond the aggregated national data of the United Nations, and thus develop firm internationally recognized characteristics to which all nations in the civilized world must adhere, at least in terms of ideals to work toward.

There will doubtless be strong ideological components based on whether a society has a free market or a central planning system, or whether it has a multiple-party or single-party political system. There will also be differences in terms of the punishment and restraint characteristic of a region or nation. But these items can themselves become an area of comparative investigation. That is to say, what nations under what circumstances are in violation of human rights when the same practices in other nations may be characterized as fully observant of such rights becomes a problem.

What are the norms that are expected in the human rights field from a nation? Here certain items can be addressed. First, nations should provide annual reports on social questions, just as they now do on economic questions; in that way attention can be focused on specific patterns of human rights violations or observances. Second, there should be encouragement of independent, non-governmental organizations that can both monitor and pressure official reports in order to gain credibility and reliability. Third, social science is uniquely equipped to monitor unusual conditions or emergency situations, such as famine, floods, and earthquakes, as well as victims in chronic misery and distress. In this way human rights reporting will not become mechanistic and so ignore flash dangers. Fourth, there can be examination of whether violation of human rights is being conducted officially or unofficially by governments, or whether governments use conduits to engage in human rights violations: this distinction, too, will prevent the monitoring of governments from becoming mechanical, which could mean ignoring the utilization of agencies perpetuating human rights violations. Fifth, insofar as possible, human rights reporting should be made uniform and should be monitored.

It is dangerously optimistic to speak of a unified human rights model based on an affluent First World and an impoverished Third World. What has been achieved in the human rights area is both more convoluted and complex than a unified world model approach would have us believe. The realities are increasing stratification at the expense of equity, and hence movement both away from and toward expanded human rights. We are witnessing a globalization of nationalism, which is displacing the internationalization of the economy. The transfer of huge chunks of real wealth to energy-rich sectors of the Third World does not alleviate the problems of human rights, since equity within each of these Third World nations remains even more elusive than it was in the past. The comparisons of wealth and poverty within each Third World nation points up the weaknesses of North-South or even East-West models. The classic gap between rich and poor, exploiter and exploited within each nation on earth is still where the human rights issue must be confronted as the living tissue of human realities.

The ambiguity of human rights as a slogan became apparent when decisions had to be made about specific nations in the Third World, and American responses to specific violations in such places. What began in the mid-1970s as an effort at embarrassing the

Soviet Union, by the early 1980s ended in self-embarrassment. Rather than point up the bankruptcy of the Helsinki accords, human rights as moral posture highlighted the continuing isolationist grip of the Vietnam trauma in American foreign policy. The lesson of the human rights episode is by no means the worthlessness of the effort, but rather its impotence at the level of moral absolutes. The efforts to implement human rights as an element of foreign policy foundered in ambiguity as a result of the sheer incapacity to develop a set of consistent guidelines. Worse, human rights as a policy goal even of pragmatic proportions came upon hard times, embittering the architects of a policy based on absolute moralism, and giving quiet, smug satisfaction to advocates of an amoral *Realpolitik*.

13

Bureaucracy, Administration, and State Power

The world today contains many societies at different levels of development that have different kinds of responsive and influential administrative networks. Although the term "pre-industrial society" has no recognizable denotive content, it has become virtually unavoidable and, if meaningfully defined, has its uses. The character of mechanization has shifted, and the role of policy making has grown, but these changes suggest not a post-industrial society but simply an evolution of the industrial order. We still inhabit a universe largely defined by pre-industrial characteristics. The industrial sector remains far from dominant in social systems.

The phrase "post-industrial society" is infelicitous: it differentiates an inherently transitional movement from something called "industrial." The phrase is ambiguous: it refers to cultural norms in the most advanced nations and is subject to a variety of interpretations, few of which are operational and even fewer mutually compatible. Societies do not emerge *de novo*. We are still trying to ascertain meanings to ascribe to industrialization. Much has happened since the end of World War II—above all, the ability of mankind to "totalize" destruction and practice genocide. Authoritarianism, no longer confined to a single-state apparatus, has expanded to the global decimation of peoples.

The term "post-industrial" is often employed as a hygienic way saying "authoritarian" in an age of the dominance of the public sector, or at least the ascendance of policy over politics. The post-industrial vision admits only of a "deeply pessimistic" view of society and conjures up an image of engineered totalitarianism. Whether such a series of postulates adds up to post-industrial society or a neo-technocratic state with authoritarian tendencies is a choice of language and representation of different levels of pessimism. But the contours of social structure and social stratification have undoubtedly shifted dramatically.

The principal characteristics of Western society are private property, private control of investment decisions, and an industrial base that primarily demands technological efficiency. Along the axis of technology we may identify developmental sequences: pre-industrial, industrial, post-industrial. There is an inherent contradiction between the principle of bureaucratization, based on hierarchy and role segmentation, and the principle of equality, based on participation. The social tensions in Western democracy have been framed by the contrary logics of bureaucratization and participation, supported by a change in institutional scale that leads to a pronounced shift in functions. The major principle of post-industrial society is the codification of theoretical knowledge and information, specifically about new technological-scientific activities based on computers, telecommunications, optics, polymers, and electronics. Control of the means of information augments the struggle over the means of production, and hence the character of work alters significantly. Work becomes a struggle among persons rather than against nature. In a post-industrial universe, society becomes a free choice by free people, rather than a banding together against nature or an involuntary banding in routinized relations imposed from outside.[1]

The theory of public administration emerged as a justification and celebration of developing practice. It has an inevitable bias toward the practical, toward getting things done, and an impatience with frustrating questions that in their breadth and scope may prove incapacitating.[2] The private woe of public administrators has been theory writ large. They have preferred middle-range theories, and, if those are unavailable, no theory at all. This is not the first time that practical people have lived a full and useful life without knowing what they are doing theoretically.

The relationship of public administration to economics, sociology, political science, and psychology bespeaks a search for broader meaning. As administration becomes distinguished from its parts, it can no longer simply make do with makeshift definitions. How does one train a public administrator for the Department of Commerce rather than for HEW? How does one train city administrators in contradistinction to national administrators? Determining entry-level skills becomes more exacting in an era of affirmative action. There are questions of the relationship between poltically appointed administrators and civil service administrators that must be answered.

If public administration is not a science, is it an art? Here the giant assumption, central to understanding government per se, is that administration is a function of state authority or authorities. One administrates things, peoples, and ideologies on behalf of an employer called the state. Administrator and state directly connect with administration and policies. The conventional nineteenth-century image of social classes that manipulate bureaucracies to gain advantage is not accurate. Rather, a big administrative machinery that manipulates other sectors is decisive: labor forces decrease in size and increase in power.

Public administration takes place in a context of policy inversion: a transformation of power from the economy to the polity, mediated by the monopoly of allocation and distribution of public funds. Public administration services not one class or sector but the state. The class interest of the middle administrators is the class interest of top administrators to the extent that they are directly linked to forms of state power. Public administration is therefore inextricably linked to state power. To say that administrators are politically neutral is naïve. To be neutral with respect to a party or an interest group is perfectly reasonable, even a prerequisite of sound administrative procedures. But for the administrator to be neutral about state power as such is a contradiction in terms. For the essential politics of the administrator is the survival, prolongation, and strengthening of the apparatus. This is administrative politics writ large, whatever the disclaimers about political participation writ small—i.e. party loyalty or partisan activity.

The essential contest in class terms occurs between elites and masses or, expressed ideologically, between statism and populism.

Those with suspicious, anti-state populist activities and those with populist concerns can no longer rest easy with old-fashioned formulations about class. Today, classes, sexes, races, and ethnic groups directly confront state authority. The legal mechanisms and financial systems make possible broad layers of social change in attitudes and behavior precisely by reinforcing the ascribed features of the status system.

Administrators do not oppose change. They demand that changes occur through a legal enlargement and acceptance of the administrative apparatus. To the extent that the system as a whole can manage a framework of rewards and punishments which appeals to the state, state power becomes enhanced and public administration enlarged. To the extent that the state cannot manage rational rewards and punishments or cannot orchestrate and mobilize the economy to do its bidding, it invites popular opposition and rebellion.

The post-industrial world resembles the post-feudal, monarchical world of Machiavelli and Hobbes more than it does that of entrepreneurial capitalism. Again, the relationship of ruled to ruler has become simplified. Mediating institutions such as parties and pressure groups are reduced to fundamental conflict and competition. In the dialectic of post-industrial conflict the manipulation of funds and forces rather than the administration of people becomes a rallying cry within administrative work, much as it did for the nineteenth-century utopians who argued for the withering away of state power.

The rationalization process compels the state to confront new leviathans, from multinational corporations to space programs such as NASA that require budgets equal to conventional government expenditures. Public administration turns to the state as a tool for monitoring and evaluating potential threats to the commonwealth. The emergence of regional trade and national associations also requires a tighter framework for state decision making. Everything from arranging sales to supervising contracts demands administrative servicing. While mediating agencies within a nation lose some of their potency, the need for state power grows as a response to external agencies.[3] The state must negotiate with a wide variety of foreign and domestic agencies and interest groups; public administration does not lack new areas to conquer.

Much is ambiguous in public administration; its size and scope remain obscure. As late as the mid-1930s in the United States, recruits for public administration came from among a wide variety of professions and disciplines. But by the postwar period public service had become a field unto itself. Administrative personnel operate in the executive and legislative branches of government; serve interest groups that range from veterans and home builders to the handicapped and police chiefs; work in regulatory and monitoring agencies; manage the roughly 500 separate congressionally sanctioned grants-in-aid programs; and deal with lobbyists. The size of the federal administration has increased dramatically.

As commitment to administrative expertise increases so does the electorate's sense of diminishing returns. These considerations pertain only to the federal bureaucracy in Washington, D.C., in which administration is the central occupation. Recent demographic estimates of state and local government administrators—exclusive of salaried managers and those who work in private administration—count nearly 5,000,000 people working as public administrators, nationwide. Between 1950 and 1979 public administration trebled.[4] The 1980 census will undoubtedly show further growth. Public administration represents a large segment of the labor force: well educated, increasingly specialized, and willing to interpret and execute the will of the state.

The post-industrial concept is organically related to public administration and its enlarged role in changing class relationships. Weber's trepidations about the growth of bureaucracy were well grounded.[5] The public sector has grown at a pace and in power far beyond older classes. The proletariat has declined as capital-intensive industry has replaced manual labor; family farms have shriveled in size and number as technology has improved food and agricultural production; the bourgeoisie as an owning group has been allowed to remain constant in numbers, but its decision making has been reduced by the need for public-sector support.

Middle management in industry has grown apace with the public sector. The incomparable decision-making functions permit one to think of industry, no less than government, as a set of bureaucratic institutions or leaders with divergent ideas.[6] But if neither government nor industry offers a single rational actor, both offer two central facets of advanced industrial society that operate with profoundly different models of economic well-being. The contrast in economic philosophies between the public and private sector is well known. The critical contradictions between these two sectors prevent any new class of administrative workers from emerging despite functional similarities.[7]

Public administrators function as a sub-class of the public sector, and industrial managers function as a sub-class of the private sector. The character of class competition also changes dramatically in the post-industrial world. The familiar competition within the private sector between bourgeoisie and proletariat yields to the less familiar but more potent struggle between the administrative vanguards of each sector. Older forms of class competition remain intact, but even these are mediated by new administrators and their authorities, interests, and regulations.

The Sherman Antitrust Act of the late nineteenth century introduced a tension between the public and private sectors that has not yet abated. It is precisely the tilt toward the public that permits one to characterize our epoch as post-industrial. Implicit is a shift not only from goods to services, or commodities to ideas, but, more exactly, a shift in the locus of power and the force of numbers.[8] Public administration is properly perceived as a sub-class that functions as an independent social force and represents the public sector in its great struggle to enlist the private sector in the goals of equity and opportunity. The public administrators function in a classic bureaucratic mold, as representative of the national interest against the industrial managers. Whether public administration can do so without undermining the innovation and initiative of an advanced society will be a challenge during the next decade.[9]

The power of American-style bureaucracy emerged with raw pragmatism that rests on problem solving. Public administration grew with only a barely disguised animus against theorists—those for whom a "rational comprehensive" model has had meaning. Public administration was celebrated as "the science of muddling through." The only widely acknowledged problem was the swiftness of technological innovation, so that there was an "absence of enough persons who are knowledgeable in computer use."[10] The solution was better functional division of technology and more extensive training.

To look back even a decade is to observe a brave new world of public administration, in which technological breakthroughs came rapidly and solutions to thorny problems were registered through improved data control. Governance through data control reached its height in the United States with the Vietnam War, the Watergate scandals, and a series of uncovered legislative deceptions. As the mechanics of administration became unmanageable, the science of muddling through led to a series of overseas defeats and domestic miscalculations. As a result, there was a dramatic contextual shift from America to Europe in the theory of public administration.

The major theses of the 1960s articulated in texts by major theorists, posed problems that have remained unresolved. Little more could be expected when even the most advanced researchers proposed that public administration overcome its "crisis of identity by trying to act as a profession without actually being one."[11] The implications were clear: concern for the social conditions of administration yielded to the professional status of administration. A review essay of the ten leading public administration texts pointed out, "Only one author devotes much space to posing the question of which groups and social class are best served by the administrative structure of the state."[12] As the emphasis has inexorably shifted from broad social issues to narrow professional issues, micro-methodology has replaced theory. Problems of measurement, evaluation, and even monitoring of programs and plans have rapidly displaced more fundamental concerns. Under such circumstances the sources of new inspiration in public administration inevitably shifted.

Innovations in public administration theory shifted to France in the 1970s. During the 1960s those who saw government in functional categories confronted those who saw government as a total system, i.e. in structural terms. The debate raged between ideologists, who preferred the language of bureaucracy because it reflected the actual hierarchy of government, and technocrats, who preferred public administration precisely because of its non-hierarchical or functional characteristics. Those who urged a general-systems perspective were denounced by those who insisted on a partial theory. Some saw bureaucracy as generic, others saw administration as generic, with public administration as one facet of it. Finally, they argued about whether public administration should be considered a separate discipline or part of the social and political sciences.

In France, after the collapse of the May 1968 student rebellion, the higher-level functionaries, disturbed by anarchism on the Left and the fascist potential of the Right, began to examine the premises of post-industrialism as a question of elitism versus populism, rather than as one of social classes or even of social functions. In the hands of public administrators the state became the guardian of public against private interest. Two schools of thought emerged: one led by Michel Crozier, who argued that the administrative apparatus had as its unique charge the supersession of social stagnation by institutional investment to render habits, negotiations, and systems of rules more complex, more open, more comprehensive, and more efficient.[13] The constraint on investment was not lack of finances or political will but need for serious analysis of change and a deeper personal involvement in institutions.

In contrast, Nicos Poulantzas, in his seminal work, announced that the decline of legislative power, the strengthening of the executive, and the political role currently assumed by the state administration now constitute the tripartite leitmotif of political studies.[14] Ernest Mandel elaborated on this theme by noting the heavy increase in the service sector and hence the expansion of administrative state power. Unlike Poulantzas, he argued that such a shift does not "lower the average organic composition of capital."[15] According to the French intellectual Left, bureaucracy should mainly be viewed not as an impediment to efficiency but as a response to structural deformities within the private and entrepreneurial sectors. The economic role of the state, not the need for efficiency or innovation, undergirds the expansion of public administration. For Poulantzas, administration is the terrain on which an "unstable equilibrium of compromises" between the power elites and the popular masses takes place and is elaborated.[16] His view is a far cry from the earlier Marxist characterization of bureaucracy as serving a class.

Crozier argued that administration has a central role in adjudicating the claims of interest groups and in bringing about innovations that cannot be made by the private sector alone.[17] Poulantzas, deeply anti-totalitarian, centered his attack on the gigantism of state power. He concluded his most recent work with a full and vigorous critique of Stalinism as an extension of Leninism.[18] But the 1970s ended with the same sad indecisiveness as the 1960s. Resolution of the issues on the basis of dependence versus the autonomy of public administration is merely a more advanced form of the same theoretical dead end that characterized American thinking in the 1960s. Researchers like Crozier are essentially policy-oriented meliorists, interested in getting society going again, whereas theorists like Poulantzas and Mandel hold no hope for moving forward through administrative efficiency so long as a market economy remains the essential motor for economic development.

The problem partially arises from the reification of administration itself. It makes little sense to argue an anti-capitalist premise when state power grows with equal force in planned as well as market economies. As Jean-Jacques Servan-Schreiber shrewdly notes, the ability of the Soviet Union and East Europe to ensure a minimum living wage to their entire population depends on this maintenance of a high degree of coercion.[19]

Today the politics of administration concerns only its own expansion; more generally, its specific political aims are circumscribed by the larger forces of the state and economy. The concentration of administrative power occurs in all socio-economic systems designed as post-industrial. One reason for employing the term "post-industrial" is that it avoids the reductionism of identifying the concentration or curtailment of power with any one kind of economic environment or party apparatus. Differences between social systems and nations have not evaporated. They remain potent cultural forces. In any democratic culture, control is in the hands of the popular sectors, which struggle against bureaucratic regimentalism. Personal leadership is common to all types of economic systems. Public administration with its impersonal ethos competes with the political apparatus it presumably serves, with its continuing emphasis on personalism and charismatic authority.

The administrative apparatus, however close it might appear to the state apparatus, must remain responsive to the led no less than the leaders—to the broad stratum of taxpayers outside the decision-making process. Public administration cannot be characterized as part of the state any more than it can be viewed as a buffer against it. They must try to mediate between the machinery of power and the interests of the public while trying to control the budget. This sort of conflict becomes an essential proprietary consideration in post-industrial environments in which older forms of conflict have disappeared. Polarization remains deeply embedded in advanced social structures but takes the form of efforts to dominate funds rather than classes.

The relationship between the planners in advanced states and the financiers (middle-management bureaucrats or public administrators) is especially intriguing.[20] It can either generate severe friction between those who make policy and set the nation on its long-range course on the one hand and those who carry out policy and hence do not easily tolerate interference or sudden shifts of direction on the other; or it can become a symbiotic relationship in which forces are joined to preserve and extend state power. Since decision-makers at the top have natural antipathies toward those in middle management, one might will expect antagonism. Increasingly, these antipathies are held in check by decision making, which has itself become post-industrial in the special sense that data determine decisions. In this evolution decision making has increasingly been removed from the political arena.

Certain characteristics of bureaucracy prevail over policy-making concerns. The growing professionalization of government work, whatever the type or level, causes a reaction. As a result, one should not expect the end of public ideologies so much as the decline of party politics. The essential danger is closure of political participation. The main question is not how administrators function in post-industrial societies but how government administration—local, state, regional, and federal—begins with a theory of negotiation between policy-makers and political officials but ends with the assumption of the reins of state power.

Politics has become identified with the national interest. The defense of those presumed interests against overseas transgressions is central. Hegel has overwhelmed Marx: the state, not the class, is now the main organizing premise. This has become as evident in developing as in advanced areas. Public administration is uniquely situated to take full advantage of the new conditions. Reared in a tradition of professionalization before politicization, organized to serve the needs of the state directly and not any one subset of special interests, and directly tied to governance without the encumbrances of elections, it is the basic sponsor of the national interest, the guardian of the state, and the arbiter of its specific survival capacities.

Hence, there is a growing conservatism within the public administration sector, not an ideologically traditional quest for community and order but a determination to preserve the state and its interests against military incursion from abroad and fiscal disintegration from within. Public administration is able to prevail because mass politics has long since been tranquilized by the narcotic of the communications media. The steady expansion of public administration, particularly below the federal level, which seems saturated, binds the administrative apparatus and further inhibits the emergence of any "grass roots," local, state, or regional opposition. A certain equilibrium longed for by federal administrators has been established.

The attempt to distinguish administration from politics is understandable, but it ultimately remains an exercise in futility. The bifurcated vision derives from a Western democratic belief that professionalism is somehow uniquely apolitical or, better, purely instrumental.[21] As Gordon Tullock puts it:

> There is no way this sort of ultimate policy formation by low-ranking personnel can be avoided; it will arise on occasion in all organizations, no matter how efficiently these are organized. . . . Not only the initial decisions, but all subsequent decision, may be made by men operating at the lower reaches of the hierarchy. The sovereign neither ratifies nor disapproves of these decisions either because the chain of command is so clogged that he does not hear of the issues at all, or because he is lazy, or because he fears that any decision on his part will, in turn, annoy his own superior in the hierarchy. In such an organization as this, the lower ranks, after perhaps vainly trying to get the higher officials to take action, may be forced to make decisions. Out of a series of such events, a sort of organizational policy may develop by precedent, and the higher officials may never have to make any choices of significance at all.[22]

In a period of computer technology, administrative decisions by fiat become frequent. The equation of expertise with administration leaves out the people. But since the people are the ultimate repository of power, elected officials are compelled to take responsibility for decision making. Opposition to this enlarged federal apparatus will come from a revived populism—from an opposition to government and its services that will undoubtedly cut across traditional Left and Right lines. Proposition 13 in California provides one indication. It is impossible to convert this populism into a Left or Right impulse, any more than one can identify government administrative actions as Left or Right. The shape of things to come is already plain: a struggle between the state, with its organized participants, and the people, with its interest groups, tax resistance, and, ultimately, sheer numerical force.

Violence, including sabotage and terror, is more likely in a non-political context than in a political one. Politics is an essential cement of moral authority, for it alone guarantees a sense of participation and control over state power. But in the antiseptic climate of top-down rules and regulations, in which party identification is low and political allegiance shaky, the legitimacy of the state can be severely threatened. The administrative machinery of government appears to be the government itself. Popular actions will become less symbolic and increasingly oriented toward direct action. Things get done by threatening bureaucrats. Hence, the web of government—that delicately laced network of authority and legitimacy—can easily break in a brave new world of bureaucratic determinism.

Post-industrial technology has been unevenly absorbed. Line managers have been afraid of the new technology, while those staff members who procure it have emphasized the efficiency of machines rather than the effectiveness of programs. As one recent report indicates: "The impact of computers on the federal government's operations and procedures is pervasive; and federal expenditures for computer-based information systems exceeded $10 billion in fiscal year 1977."[23] In the immediate future much more effort will be devoted to developing a symbiosis between electronic information systems and administrative tasks; it will be necessary to model complex systems, as well as to process routine data. Forecasting and assessment of technology are linked to a greater use of advanced computer systems. Until now public administrators have focused on short-term uses of advanced technology. In the future long-term planning is likely to prevail. The danger is that complex decision making will become the responsibility of machines rather than human political actors. Here is an area of great potential danger for public sector policy making.

However moot the question of the "good" or "evil" potential of technology is, public administrators not only react to a post-industrial technology but have long helped guide it. The military and nuclear sectors are particularly vulnerable to the bureaucratic management of technology and science. Directing technology involves two kinds of apparently contradictory activity which are also opposite sides of the same coin: the encouragement of technology and its regulation.[24] Because of the enormous costs in research and development and the consequent impact of profitability, technology has become an organic part of public administration and decision making.

Administration is a service as well as a system. Advanced societies evolve toward increases in the size of this service sector. But against this inexorable tidal wave of paperwork, allocation, and decision making stands the spector of a new public managerialism—of a new class that loses intimate contact with ordinary events and people and so jeopardizes the political system as a whole. Even if bureaucratic administration is impervious to internal assault, its removal from the taproots of incentive and innovation make any society susceptible to destruction from without.[25] The need to consider public administration as a servant or at least an agent for developing incentive, and not a master of government, is the alpha and omega of democratic life. It also ensures the future of public administration as an entity apart from special or private interests.

It is dangerous to conceive of post-industrial technology as necessarily feeding the fires of administrative domination. Mindless slogans and irrational specializations tend to thwart any sense of citizen participation. But certain developments like cybernetic systems design can facilitate radical, technological, and social decentralization, no less than trends toward consolidation. The distinction drawn by Manfred Stanley between technology and technicism is useful. The aims of education, whether administrative or political, are serviced by the redesign of social research methods, the institutionalization of linguistic accountability, and the criticism of irrational specialization.[26] The pace of technological innovation has quickened and the size of bureaucratic administration has grown. On the assumption that historical clocks cannot be turned back, the need to harness the former and reduce the size of the latter is a major task of Western culture, which must itself be made sensitive to new patterns of administration in newly emerging parts of the world.

Conflict between branches of government will remain in a technological context. The executive branch will still mobilize its bureaucrats to disprove the claims of legislatively appointed bureaucrats. But such conflict among administrators should become less significant as the tasks of governance become more complex. The entire arsenal of populist politics will be required to offset the advantages of bureaucracy. Electoral politics are episodic, sporadic, and ideological. Administration is continuous, fluid, and rationalistic. If the two sides need each other, in a more proximate sense, such a need is tempered by recognition of different constituencies, functional prerequisites, and, ultimately, different visions of law and order. Maintaining unity through opposition, is an essential task for constitutional law and an important function of the modern judiciary. Democracy has been and will continue to be affected by the rise of public administration in a post-industrial society in which state power grows and in which the relationship between politics and economics is inverted, with politics becoming the base and economics the superstructure.[27] Democracy will also be attacked for its inefficiencies. That is the essential meaning of totalitarian post-industrialism. The adjudication of the innate tensions of mass demands and organizational rules is the essential role of public administration in a democratic community.

Democracy is definable decreasingly as a system and increasingly as a process of mutual articulation of interests that can be adjudicated and negotiated in an orderly and non-injurious way.[28] Conflicts between public administration and mass politics should be viewed only as a late twentieth-, perhaps a twenty-first-century possibility. As old conflicts between judicial, legislative, and executive branches of government decrease in importance through the increase in the technology of scientific decision making, a clustering effect may occur in which a government that routinely interferes with individual life is pitted against one that leaves people alone. The private sector will come to mean those areas of individual performance left intact, without special interest incursions, rather than simply a portion of the economy left under entrepreneurial control.

A new system of checks and balances might evolve in accord with the emerging characteristics of the political process as a whole. The basis for such programmed differentiation of tasks and distinction in goals may not be spelled out constitutionally except as a series of mandates that permit administrative functions in a pluralistic context. The specific functions of various branches of government will appear more rational in a context of opposition or at least of functional differentiation of administrative-bureaucratic tasks. At one level this is the politics of the last word. The executive who spends twenty hours on the budget has the final say, not the administrators who spend thousands of hours in preparing it. The congressman or originator, purveyor, and mediator of preferences must vote on legislation, even though bureaucrats and managers inundate him with evidence that a bill is too costly or not operational.

Major movements of our time revolve around the size and strength of the administrative bureaucracy. It is necessary to make it responsible, increase its efficiency, and decrease its power. To the extent that the obstacle to further political development is administration itself, either through tax pressure or through withholding of political legitimacy—to the extent that the obstacle is the individualistic propensity of the popular will—one can expect political elites to prevail over popular forces.

Whatever the nature of the economic system, our epoch bears witness to a constant expansion in state power, bureaucracy, and administrative domination and disposition of people. The economic system a nation lives under has become less important than the fact of state growth and its allocative mechanisms. New forms of popular rebellion have little to do with class membership and cross traditional social boundaries. Advanced technology either ensures a simplified bureaucracy more directly responsible to the people's expression of choice and interests; or is a mechanism to impose decisions from above. The new technology offers potentials for both dictatorship and democracy. The character of state power uniquely determines how these potentials are realized. The outcomes are in doubt, not because of the feebleness of social research, but because of the willfulness of human actors. Before the torrent of populism the wall of every known elitism threatens to crumble. But populism itself provides a new form of irrational solidarity—a tactical option to elitism that still leaves undisturbed the essential structural inequalities of advanced post-industrial societies, East and West.

14

Social Welfare, State Power, and Limits to Equity

American society presents a major inconsistency: a near-unanimous belief in the value of equality and a persistence of vast income and occupational differences. The leveling impact of mass communications led to a corresponding decline in an industrial working class, further accelerated by a rise in service industries and a fall in productive activities. The gap between values and interests, while not any larger than it was in the past, has been viewed as more intolerable. Hence, much more pressing demands for equity practices and legal safeguards have now been registered, along with the termination of all forms of inequity, to the extent that this is possible.

The State as Guarantor of Equity

Before taking up subjects of the limits of equity, it is important to register how the rugged individualist form of the competitive society has broken down. First, many Americans have begun to realize that the limit of a geographical frontier has finally been reached, and that, consequently, the corresponding ideology of striking it rich through land acquisition has come to an end, being reduced to home owning. Second, as society itself has become more politicized, the population has become less patient with inequality and more aware of the bureaucratic and corporate obstacles that contribute to a feeling of powerlessness. Finally, the affluence stimulated by the end of World War II enabled many Americans to raise their incomes to a point where they were no longer concerned just with making ends meet. As a result, new expectations emerged, not only for a higher standard of living but also for gains in the quality of life and in the control of one's own destiny.

In addition to these precipitating factors of equity demands, there have been structural changes within American society that give force and urgency to the concept of equity. Foremost of these structural changes are the size, role, and legitimation of state power. The American government has grown from a small organization to a large, complex network of organizations, and most of this growth has taken place in the twentieth century. The very rapidity of bureaucratic growth contributed to demands for equity since hiring practices, social relations, and ideological demands are all made in terms of the federalist credos rather than in terms of the old capitalist ethic with its implicit neo-Darwinian struggles for survival.

Growth and size alone do not explain the structural basis of equity demands. The changing role of the state, connected as it is to the changing base of legitimation, must also be taken into account.[1] The major role of the federal government in the nineteenth century was to provide a favorable environment for business activity. The state, within this framework, was Hamiltonian, involved with the stabilization of currency and the steady legal pattern that ensured the rights of exchange in contracts. In addition to the support for business activity, government came to provide for services and facilities in economic areas that individuals and corporations could not provide for themselves. Hence, everything from national roadways and waterways to the development of the National Aeronautics and Space Administration required government support in order to be remotely successful. The idea of government support at this level was still directly linked to commerce, or better, to those areas in which the public sector could perform with far greater rationality than could the private sector. This type of legitimation for federal involvement in the affairs of society might be viewed at a legal, rational level. It ensures that the state provides for a stable and predictable environment for the growth of commerce and capital. The government legitimated the expansion of its own state power by presenting its activities as a favorable approach in support of the economic system. Calvin Coolidge put the matter bluntly when he said that "the business of government is business." This was carried further by Charles Wilson when he declared that "what is good for General Motors is good for America." The state, in other words, grew by assisting the business and commercial activities of society. Anti-trust legislation was first intended to check the evils of monopoly in order to make sure business activities flowered.

In American society as in other advanced post-industrial systems, the state began to take on a role quite apart from assistance to business and commerce. Anti-trust legislation, for example, which began by checking the evils of monopoly, ended by checking the evils of commerce itself. The state changed its role slowly but surely from defending entrepreneurial interests to creating some sort of balance between classes that would permit no one class to rule. By default, if for no other reason, the change of role made the state become even more powerful, albeit different in character, in the late twentieth century from that during the previous stages of capitalist and socialist development.

With the Depression and the New Deal the role of government in the United States (as well as other industrial societies) began to change dramatically. In their new guise governments sought to improve not only business conditions but the condition of the population as a whole, even if this meant a curb on profits. The state began to intervene more directly into the economy for the purpose of creating upward mobility and greater participation, thereby ensuring that the stratification system would not harden. Some of the early activities in this area involved the recognition of labor unions as co-partners with business associations in the conduct of commerce. This was done through a series of legislative measures favorable to union activities. An indicator of this changing attitude toward equity was the growth of social welfare programs, which in the United States began during the New Deal and was accelerated under the Fair Deal, the New Frontier, and the Great Society. All of these actions were presumed necessary, not because individuals failed to take advantage of opportunities in the capitalist system but as a consequent belief that the bour-

nomy made equity as such impossible. In this way, the ..lated itself not only in terms of business needs but in ..ms of improving the status of all groups.

In the process of presenting its goals in these terms, the state inevitably made new demands far beyond the economic realm envisaged by policy-makers in the New Deal "Brain Trust." The state inexorably and increasingly become involved not simply in matters of economic equity but in those of racial, ethnic, and sexual parity. The state presented itself, and was perceived by the wider public, as the source of law, and hence the ultimate recourse for those who had legitimate demands. No longer was the archetypical struggle waged between workers and bosses: it was between interest groups mediated by the force of law embodied in the power of the state.

The civil rights movement of the 1960s can be considered important in this general revitalization of equity issues in two fundamental ways: first, the methods of the movement involved a direct appeal to law and the government for the restitution of civil rights; and second, these demands differed from the demands of earlier periods. Methods used by the black movement differed considerably from those of the trade-union movement in that demands were directed more to the government, particularly on the federal level, than to entrepreneurial groups and corporations. In this way, the black movement attempted to bring the federal government and state power directly into play to ensure equity. Whatever it took (federal troops in Little Rock, or Supreme Court decisions in Washington), there was the belief that the government had to support demands for equity or be revealed as ideologically bankrupt.

The second factor in the civil rights movement involved a sense of rising expectations that again could only be assuaged by federal support and intervention. Early victories by blacks in creating a raised level of educational and cultural institutions and equal access to public accommodations further stimulated a desire for equity. This was followed quickly by demands for similar pay for similar work and for open housing; by voting rights pressures that led to the election of a great number of black public officials; by community control; and by affirmative action, which one might say led to a change of emphasis, since what started as a demand for equity terminated as a demand for pluralism, or the right to be different. Separate but equal once again became a slogan this time at the urgings of black rather than white groups.

The utilization of the state by the blacks as a last resort led in quick succession to the same approach by women and by ethnics. The leadership of the women's rights movement reveals a formal similarity to the black model. First, there is the older model, which anticipates a society in which minority groups are ultimately absorbed into the mainstream by losing their distinguishing characteristics and acquiring the language, occupational skills, and life-styles of the majority, which does not vanish but remains a *leitmotif*. Second, there is the pluralist model which anticipates a society in which racial, religious, and ethnic differences are retained and valued for their diversity, yielding a heterogeneous society in which it is hoped that cultural strength is increased by the diverse strands making up the entire society. Finally, a hybrid model emerges, in which there is a change in both the ascendant group and the minority groups so that changes take place not only among blacks, women, and ethnics but among the dominant racial, sexual, and religious groups.

Although the demand for equity is dominant within vast social movements, the curious fact is that the government, far from being feared as an opponent, is strongly encouraged to intervene. The assumption of "oppressed" groups is that legitimate or judicial review or lobbying are likely to lead to equity rather than more repression. This is clearly a departure from the traditional view of the state as a supporter either of the *status quo* or of the business community.

The old view of the state is that its role and function are in the service of privileged and wealthy classes. The new view of the state is that its role is to blunt, if not eliminate, differences of class and status. The distribution and the quality of public services affect the absolute and relative well-being of individuals. Although inconsistencies exist between classes with respect to income and resources, the state is charged with the task of minimizing such inconsistencies. The whole gamut of social services and health, education, and welfare is viewed as the federal "hidden hand," intervening on the side of the poor and the downtrodden to create an equal starting point for all concerned, whatever different outcomes may occur. The state becomes a service organization; whereas political power becomes the measure of permissible levels of inequality and freedom.

What began as a non-ideological effort to increase the efficiency and rationality of governance terminates in a new ideological concept and struggle: the state as exclusive guarantor of the democratic premise. Samuel Huntington has referred to the new role of the state as "democratic distemper." His definition of what the 1960s brought about stands as a succinct reminder that the effort to end ideology actually resulted in the re-emergence of new and more vigorous status ideology.

> The 1960s also saw a marked upswing in other forms of citizen participation, in the form of marches, demonstrations, protest movements, and "cause" organizations (such as Common Cause, Nader groups, environmental groups). The expansion of participation throughout society was reflected in the markedly higher levels of self-consciousness on the parts of blacks, Indians, Chicanos, white ethnic groups, students, and women, all of whom became mobilized and organized in new ways to achieve what they considered to be their appropriate share of the "action" and its rewards. In a similar vein, there was a marked expansion of white-collar unionism and of the readiness and willingness of clerical, technical, and professional employees in public and private bureaucracies to assert themselves and to secure protection for their rights and privileges. In related fashion, the 1960s also saw a reassertion of equality as a goal in social, economic, and political life. The meaning of equality and the means of achieving it became central subjects of debate in intellectual and policy-oriented circles. This intellectual concern over equality did not easily transmit itself into widespread reduction of inequality in society. But the dominant thrust in political and social action was clearly in that direction.[2]

So much has the role of the state as guarantor of equity become central to a new ideology that the raging debates of the 1970s involve precisely the role of federal intervention and state power in the affairs of the citizenry. The new conservatism holds that further centralizing authority, far from ensuring equity, leads to disorder and a breakdown of community. The government becomes the only center of gravity between the individual and the monolith. In its more advanced versions, the public sector itself becomes subject to ideological attack. Thus, the commercial sector, which at the beginning of the century was serviced by the state, has come to view the state as the enemy of its class interests. Private enterprise has developed a strong defensive posture with respect to any furtherance of state power. The bourgeois state of the nineteenth century has turned into the anti-bourgeois state of the twentieth century. With this phenomenon, the whole notion of the role of government in relation to the economy becomes a subject for profound review.

Weakening of the Public Sector: Return to Malthusianism

The Stevensonian call for a "revolution of rising expectations" which began in the mid-1950s is now resolving itself in the 1980s as an equally urgent appeal for a "revolution of falling expectations." Universal boosterism has yielded to particularistic pessimism. Now the cry is for "the end of progress."[3] Time does indeed march on but in directions so diverse and disarrayed that charting the marchers, much less participating in the events, has become a major undertaking whose outcome is highly uncertain. Nothing is more dangerous than to presume that the models of the world that we create are either more perfect or less elegant than the imperfections we find about us in nature and society alike.

This observation is prompted by dismay at how thoroughly unprepared the social sciences were for the so-called energy crisis of the 1970s. Social scientists were even less prepared to cope with the consequences of this new state of affairs. American social science has grown up in a neo-Keynesian (or better, Galbraithian) world of abundance. The only admissible problems faced by the technostructure concern the allocation of resources, not their availability. But the hoary demons of neo-Malthusianism have struck back, and the shared contempt that orthodox Keynesians and unorthodox Marxists alike held for the Malthusian prediction of war, famine, and plague as a "solution" to poverty and overpopulation has turned upon them.

Industrial societies in general, and American society in particular, have fed the hopes and illusions of millions concerning the royal road to upward social mobility. Equity is a right guaranteed to all by democratic constitutions, but it is underwritten by an economic apparatus that from nursery school to graduate school assumes that the path to success is work, occupation, and specialization. The road to failure is departing from this well-trodden formula. At the present time, many themes in American culture have conspired to shake loose inherited enlightenment assumptions that social welfare and economic mobility march into the future in step and hand in hand. Work and welfare have become opposite expressions of the American ethos. Energy and equity have both been subject to cost-benefit analysis that all but cancels out moral visions of the good society. We are confronted by new forms of social relations and hence new forms of social antagonisms. This chapter attempts to come to terms with what is new and old in society and by extension what is living and dying in social science.

The first point that struck a rebellious chord about the late 1970s is that in the ebb and flow of cost-benefit analysis the tilt clearly favored costs (and work) rather than benefits (and welfare). The concern is currently with the economy as a whole, and not simply with federal budgetary processes.[4] There was a precipitous shift away from a sociological and social-welfare orientation toward what might be termed an econometric and accountability model. A pervasive facet of the United States in the 1980s is that systems of entrepreneurial measurement have been established to evaluate the cost and quality of government performance. Questions are raised about productivity of labor and only secondarily about welfare benefits for those who do not toil. This represents a profound decennial shift in American society. There is a more sober and modest estimate of potentials for growth through public or governmental sectors, and a marked return to potentials for growth through production or private sector methods.

What might be considered a derivative of this changed American view is the further recognition that the 1970s was a decade of finite resources. In several dramatic forms, heightened to a fine point by the domestic energy crisis and foreign oil embargo, a new awareness has taken root that resources are limited and therefore growth is limited. This is no mere ideological preference for a zero-growth model in either the economy or demography but a response to the limits of natural wealth of any country. Beyond that, the "dependency model" applies not only to the developing nations of the Third World but also to the more developed nations of the capitalist and socialist worlds.

No less than a decade ago reports were being issued by leading economists at the Hudson Institute that by 1980 Japan would be the world's first or second capitalist industrial power.[5] There was slender awareness amongst most forecasters of the major problems of scarce resources, the more than merely routine problems of allocating abundance. Our economic obtuseness largely derived from an inherited worldview, both Marxian and Keynesian, or sometimes a combination of the two, which persists in assuming that the central problem is product allocation rather than resource availability. We are living through the shock of a change in economic paradigms in which resource availability is once again significant in our collective cogitations. We have come to rediscover a Malthusian-Ricardian world in which famine, pestilence, shortages, and subsistence are viewed as mechanisms to stabilize and rationalize economic systems through gradual population reduction, if not through outright military competition for such scarce resources.

These developments represent the end of what might be called the "optative mood" Americans have imbibed. Progress has become problematic. "Built-in controls," "the search for control in the face of license," and "prudent restraint" are now familiar refrains cautioning those who simply identify progress with freedom.[6] The exceedingly strong pessimism that has resulted is, in part at least, a recognition that resources are finite. This has also produced a growing polarization within each of the dominant structures of the world economic system.

Within the capitalist system the contradictions between the United States and Western Europe have become intense, heightened by different political and military strategies to gain necessary resources for continued high industrial and consumptive growth. Within the socialist bloc there is intense rivalry between the Soviet Union and the People's Republic of China, propelled by similar drives to maximize production output. There is a growing divergence between Third World nations with the oil and food resources that allow them to penetrate the advanced economies, at least as junior partners, and those nations in a Fourth World that remain poor in both food and energy resources. There is a geo-

.ic component to this Malthusian outlook that intensifies the .ization and breakup of world systems and empires. We wit- .. ; a divergence not only between colonies and empires but between empires and nations that presumably share similar economic and cultural systems.

The redistribution of wealth leads to a different image of the world. In the past it took international warfare to bring about major redistribution of wealth. World War I and World War II both produced dramatic shifts in power and economy but did so among those established European and Asian nations that by the turn of the twentieth century had already achieved high levels of productivity. While warfare remains a constant problem, armed conflict has mainly been conducted at sub-national and regional levels. Hence, the process of redistribution of wealth has changed from a military to a diplomatic struggle. Game-oriented decisions have been employed to bring about wider redistribution of wealth, with a more telling effect than the moral arguments of earlier decades had.

Nations such as those of the Middle East, as well as some nations in Latin America and Asia, have managed to participate in the market economy as recipients no less than as donors of real wealth. As a result, there is a shrinkage of resources among the wealthy industrial countries. Making do with less has become a way of life in Western Europe, and now North America is compelled to follow suit—against its cultural will, as it were. New slogans bombard us on all sides: small becomes beautiful as growth is limited and resources finite. Instead of an unreflective appeal to giganticism, there is the emergence of a miniaturization process most notable in relation to automobiles, housing, and other aspects of social life in the advanced nations. A new aesthetic of conserving has emerged to rationalize scarcities.

Changing attitudes toward size and greater utilization of commodities already available are functions of the redistribution of wealth. Such redistribution occurs without the warfare usually accompanying massive changes in ownership of wealth. As a result of these tendencies within the American economy, there have been corresponding rebellions against big government—which is reputed to be the source of inflated budgets and systems of welfare established without a recognition of the high cost factors involved. More specifically, the charge is made that a level of government expenditure has been reached that assumed continuing levels of economic growth that can no longer be either sustained or realistically forecast.

An essential way in which this redistribution process has manifested itself is the sharpening gap between private enterprise and the public sector. Although the private sector has recovered from the recession of 1973–75, the public sector clearly has not. As a result, those involved in the public sector have continued to feel the burden of the 1970s recession well into the 1980s—perhaps in fuller measure now than when it was under way. Almost every major indicator of economic growth—housing, new private starts, consumer durable expenditures, number of people employed, corporate profits after taxes—would indicate that the recovery of the private sector has been robust, whereas the illness of the public sector has remained severe, even chronic.[7] In consequence, the politics of this period reflect a genuine belief that if a choice has to be made between lower profitability and continuing measures to ensure the welfare of the citizenry, it is the latter that has to be sacrificed.

What lends weight to this decision of the private sector elite to weaken the public sector is its support by the group employed in the private sector, particularly the trade union movement. What sometimes is euphemistically referred to as a taxpayer's rebellion is in fact a middle- and working-class discontent with a system of federal expenditures that equalizes the high end of the welfare package and the low end of the working package. This rebellion, whether it takes the form of a taxpayer's rebellion or trade union discontent with government schemes for increasing the welfare package, reflects itself in the virtually instinctual patterns of negative voting with respect to any increase in taxation that would raise levels of federal expenditures or bureaucratic management. As a result, various programs of affirmative action, whether in education or other areas of the public sector, come hard upon tenure and seniority systems that cannot be overcome in a period of relative economic stagnation.

Such changes have served to accommodate the new era of

energy shortages with a minimal disruption to the private sector. They may have created the conditions for new types of upward mobility in the form of demands for higher levels of participation in the work force and lower levels of tax payments to the welfare rolls. There is no mistaking the harshness and bluntness of this antagonism: neo-Malthusianism joins forces with neo-Darwinianism to satisfy the claims of the working and middle classes against marginal underclasses, characterized by their dependence on government welfare supports. Work versus welfare becomes more critical than inherited class and race antagonisms in the scarcity society.

Paul Neurath recently called attention to this return of Malthusianism and predicted a continuation of energy and food shortages.[8] His own call for a serious position somewhere between pessimism and optimism, while well taken, remains to be worked out in practice. For the issues are quantitative at one level—how much of a reduction in living standards can advanced social sectors accept?—and qualitative at another—what amount of growth is an American public willing to sacrifice to retain standards of liberty and equity?

Limited Energy Resources and Equity Demands

The velocity with which new realities have emerged can perhaps be gauged by the fact that only a decade ago the phrase "Third World" was considered an innovation and an idiosyncracy. It has now become a commonplace, and in fact a new paradigm has come into being: the "Fourth World." Now a nation may be rich in energy and rich in food; rich in energy and poor in food; poor in energy and rich in food; or, unfortunately, poor in energy and poor in food. While the United States numbers itself among the most fortunate nations on earth in the first category and still very much the center of the First World, it too has had to recognize that resources are finite and that problems of allocation are nastier and more brutish when one allocates shortages.

Simple aggregate data themselves reveal the implausibility of continuing the present circumstances into the future. In the First World, some 20 percent of the world's inhabitants directly control

10 percent of the world's resources.[9] Under such circumstances, the question of diminishing natural resources must take on a grim aspect even for the most affluent country in the world. Especially telling is the fact that, although resources are reduced, or withheld, as in the case of oil, American expectations continue unabated, and interest-group politics has become the operational guideline for all citizens on the margins of society. Appeals for government relief of inequities continue at a heated pace. Statistical measures of inequality—whether by blacks in relation to life expectations, women in relation to occupational mobility, youth in relation to quality of life, the aged in relation to security and health measures, earning gaps between different regions of the nation—are often pressed by sub-groups and coalitions. Whenever resources shrink, demands increase. The ideology of equality has become more pronounced as the capacity of American society to fulfill demands for equality has become strained.

It would be completely erroneous to assume that such demands for equity are without merit. A report issued by the Council of Economic Advisers[10] noted that the bottom 20 percent of all families had 5.1 percent of the nation's income in 1947 and almost the same amount, 5.4 percent, in 1972. At the top of the economic ladder there was a similar absence of significant change. The richest 20 percent had 43.3 percent of the income in 1947 and 41.4 percent in 1972. Thus, while the incremental wealth of Americans continued to rise during the post-World War II economic cycle, the ratio of wealth to poverty hardly budged. This is not to deny that notable gains were made on a sectoral basis: for example, the median income of black families went from 57 percent of that of white families in 1959 to 76 percent in 1972. Yet people who are defined as poor are now poorer in absolute terms when compared with the rest of the American population. In 1959, those defined as poor had about half as much income as the average family; in 1972, they had only a little more than one-third as much. And these statistics were compiled prior to the recession that shook the United States between 1973 and 1975.

Even were we to accept the premise that American society can fulfill the main prerequisites of the drive toward equality for large population clusters such as working women, black males, or minorities whose native language is not English, there remains a darker side to equity demands. They are unending and ubiquitous. As David Donnison reminds us:

> There are in fact so many different patterns of inequity in a complex urban society that to call for more (or less) equality without specifying which pattern concerns you is a pretty vacuous appeal. The main patterns are as follows: *The lifetime cycle of incomes* producing for all social classes, successive periods of relative poverty and affluence during childhood, early adult life, early parenthood, middle age, and retirement. *Spatial inequalities*, due to (a) interregional, (b) urban-rural and (c) intra-urban differences in opportunities and living standards. The patterns of *social stratification* within urban, industrial, bureaucratic societies which produce social classes with differing bargaining strengths and differing inheritances (material, cultural, physical and intellectual). Social *discrimination* which benefits or handicaps particular groups on grounds of sex, religion, ethnic origin, accent or other characteristics.[11]

The demands for parity by short people, fat people, handicapped people, or whomever have all become increasingly plugged into a general interest-group model that American society has substituted for older class models. Demands by black psychiatrists stimulate counter-demands by Italian psychiatrists. Tolerance for Cuban immigrants was met by demands for parallel treatment for Haitian immigrants. Research into cancer brings forth appeals for basic research into the causes of diabetes. All of these demands are prefectly reasonable; none save those who are without human compassion could deny the legitimacy of such demands. All utilize the juridical-pluralist model that American society has made the test for political participation, namely, articulating interests in a legal manner and withholding political support if necessary to achieve such interest. It is only when pluralism degenerates into self-interest and threatens national survival that such interest groups themselves are subject to curbs.

When one measures this cacophony of equity demands against the shrinking resources of the 1980s, the problem of the American commonwealth as it moves toward the twenty-first century becomes increasingly apparent. For it is no longer reasonable to expect other nations to permit the United States to resolve its own domestic

tensions and prevail in the world at large with reso
by them. The demands for a New Economic Order
seriously, not as moral outrage but rather as political l
if one could muster a philosophical argument along
Rudyard Kipling's, that along with the white man's burden came the white man's prerequisites, it would be fruitless because no one out there is listening any more; there are few whites who would care to make such arguments. Thus, the deadly equation of shrinking resources and rising expectations must unquestionably meet in a head-on collision that will test the mettle of American society as never before in a peacetime context.

Addressing himself to the central issue, the relationship between continued inequality and limited growth, Karl W. Deutsch asks, somewhat rhetorically: "What is the probable effect if the world is now told to expect more scarcity, not less, and not for a short period but for a long one, and perhaps even impermanence?" His answer essentially is that a revolution of falling expectations "risks a new age of international conflicts that in the end may prove fatal to all of us." Indeed, he urges a prevention of such "a drift toward catastrophe." Among the mechanisms he recommends are the following:

> National and international stockpiling, an international system of reserves of food and fuel, the opening up of new agricultural acreage and mineral deposits, the improvement of technologies, the development of substitute materials and energy sources, the transition to less heavy but more sophisticated equipment (e.g., to transistors and printed electronic microcircuits)—all these may help to stave off a "revolution of falling expectations" and thus to buy more time for mankind to become truly joined "for better or worse, for richer or poorer" in the unity of the human race.[12]

These are by all odds meliorative measures, which at best "buy time" to stave off disaster by rationalizing available resources, and imply no alteration in basic current inequalities. The question remains: Precisely why is it that a revolution in falling expectations cannot or even ought not to take place? Indeed, one must argue that, contrary to Deutsch's views, persistent and increasing inflation and unemployment within industrialized nations serve pre-

..isely to accelerate a revolution of falling, or at least stable, expectations, and provide mechanisms for accommodation to lower or "sustainable" standards of living that will ensure the survival of the political and social system as a whole.

Early efforts at futurology, based as they were on mechanistic frameworks and models that simply assumed the continuance and extension of current ratios of international resources, make no sense. If social research proved so inept at anticipating the current energy crisis, even though the crisis was initiated as a boycott inspired by political considerations, what is one to make of research estimates predicated on events a quarter-century in the future? To deal with the future implies an understanding of the present, and that also requires a sense of how current problems can be resolved with currently available techniques. It is more than tautology to assert that problems of resources can only be resolved at this point by technology, whether the prevailing technologies—harnessing atomic energy for industrial use—or a counter-cultural, counter-technology based on harnessing bigger and better windmills. Energy resources and their discovery are, after all, problems for physical science, engineering, and technology generally. At this level we are dealing not so much with the possibility of expanding resources as with a timetable for that expansion that realistically continues to satisfy rising social expectations. If the problem of technology concerns the allocation of resources, the problem of bureaucracy concerns the allocation of scarce resources. The bureaucratic prerequisites of the moment are to harness available resources and to allocate them in such a way as to prevent an explosive civil war, a race war, a class war, or warfare generally. Thus, in point of fact, the bureaucratic problem has little to do with the limits of growth because to talk in such terms is simply to freeze the present inequities into the social system.

Those urging a position based upon the "limits of growth," such as Jay W. Forrester, have argued that the question is not, "Can science remove the physical limits to growth?" Rather, we should ask, "Do we want science to remove the physical limits?" Forrester observes that to argue in favor of continued growth "is equivalent to saying we want growth to be arrested by social stress alone."[13] Nowhere in this position is the thought entertained that corking scientific and technological mechanisms of expansion might actually cause social stress. In this framework, stress is caused by growth; whereas tranquillity is ensured through stagnation. But surely one might at least raise serious objections to this reversal of independent and dependent variables, this unexamined assumption that growth causes stress, rather than that social stress may result from a technological slowdown. The recognition that stress may increase from a zero-growth policy has led to the emergence of a middle-ground view based on slow growth. But the question still remains, How slow should growth be, and who must slow down most and/or least?

To freeze the developmental process at this time is intolerable for all marginal groups currently making their equity demands upon society as a whole. The object in bureaucratic terms, and in human terms as well, is not the freezing of growth at present levels to create a new stasis but rather the reallocating of whatever wealth and resources are available, whatever their absolute size, so that the pie is divided in a more equitable manner. The American people have proved quite capable of accepting a smaller pie. They are probably not capable of accepting a smaller pie unless the present ratio between haves and have-nots changes. The present imbalance of earnings and incomes characteristic of classes in twentieth-century United States as a whole will only exacerbate questions of redistributive justice, since the total pie will clearly be reduced by virtue of international factors beyond the control even of the wealthy. The data on redistribution that would open the stratification and participation networks are not encouraging to those who want both economic change and political order.

To enter a world of technology and bureaucracy is also to face a world of technocrats and bureaucrats. It is to leave behind old formulas based on class struggle along conventional lines of bourgeois versus proletarian. Old classes attached to economic production shrink, whereas new classes attached directly to the state apparatus grow exponentially. Nor do technocrats and bureaucrats simply grow; they *become* each other, not merely like each other. They often represent interchangeable parts in a commodities culture that serves to keep the system intact. This introduces a more advanced problem, in addition to that of responding to a world of shrinking resources and rising expectations: the success of these tasks may imply a curb, if not an end, to the democratic political and social structure that Americans have been used to. As unpalatable as Samuel Huntington's thesis might appear, his challenge is at the least one that has to be met with candor rather than with rancor. "The vulnerability of democratic government in the United States thus comes not primarily from external threats, though such threats are real, nor from internal subversion from the left or right, although both possibilities could exist, but rather from the internal dynamics of democracy itself in a highly educated, mobilized, and participant society."[14]

Whatever else democracy is, whatever arguments can be mustered in its behalf (and there are many), it is an expensive system. It involves a great deal of deliberation, competition, conflict over goals and methods, and making decisions based upon humane considerations that are more political than economical. And such decisions require high growth to cover new demands.

Here we come to the ultimate problem of the 1980s and beyond. The issue is not only the survival of the Republic but its survival in terms of a democratic framework that has come to be viewed as an extraordinary, unafforable luxury in many other societies. It may well be that the allocation system will work its magic ways. Technocrats and bureaucrats alike will attempt to solve problems of shrinking resources and rising expectations respectively. However, the engineering model of innovation works too often at the cost of democratic politics, by resorting to a policy-making state apparatus that is effective, efficient, and highly centralized; but only in short-run terms. We may be faced with a dilemma: on the one hand, to save the Republic and lose its democratic essence, on the other to continue preserving democracy and seriously jeopardize the Republic as it now exists.

The choice between the state as guarantor of equality and the community as guarantor of liberty is hardly a pleasant one. But it is a measure of the lines of future struggle that issues and institutions are perceived in this trans-class manner. Similar polarities are taking place in the world at large. Those who urge world peace, for example, often do so at the cost of continued world development; on the other hand, those urging world development often do so without much regard for tranquillity of the world as a whole. The proliferation of atomic capability is clearly the most important and apparent of these dichotomies. Whatever the resolution, we can expect that the new revolution in American society will be one of falling, not rising expectations. Or, put in a more optimistic manner, as the United States becomes part of the world community, it must share the burdens of others. It must become more like the rest of the world in its lifestyle, consumption levels, and in the productivity of its citizens. This is both for better and for worse.

It might well be that more sophisticated technology is related to decreases in productivity; and that the current malaise in worker output, variously described as alienation and anomie in the work force, is little more than an early warning that advanced industrial societies such as the United States are indeed drawing closer to the rest of the world. But there is a larger sense in which this growing similarity has dire potential consequences. At the level of political organization, only a few countries in the world—two dozen at the most—can still manage the luxury of democracy. To become part of the larger world may be to lose the luxury of democracy as well, unless it can be demonstrated once and for all that democracy and development go together, and not at the expense of the permanently poor. There are no cheap victories, no visions for a year 2000 that can spare us the tragedy of choice.

When one views future international relations as characterized by options based on interdependence, independence, or isolation, perhaps the most persuasive, if most evasive, solution is to choose all of them. The question at all times is the mix. Beyond that, whether such relationships are based on superordination, equality, or subordination depends on the national, sub-national, or supernational units to which we refer. My own position is that the demands for economic equity and social parity that have pushed their way forward within the United States are now at work at the international level. Every person and each nation must take seriously the idea that all people (and all peoples) are created equal, or risk the perils of rebellion. The concomitant approach that nations, no less than people, can only count as one will be a harder lesson for the powerful of the earth to absorb.

Equity organizes the relationships of the smallest national units with the largest national units: Albania with the Soviet Union, Puerto Rico with the United States; and, for that matter, Nicaragua with Mexico. "Sovereignty," like "the person" in Anglo-Saxon jurisprudence, is a legal entity. It demands liberty in relationships based on rather powerful constraints of law. This serves to underwrite and underscore a continuing tension between nationalism and individualism. Equity demands compel a powerful drive toward various types of redistributive mechanisms that may prove to be undemocratic, a drive that certainly moves counter to traditionalist concepts of liberty. The dealings of nations with each other have increasingly been characterized by a cautionary but positive spirit. These impulses may in part be thwarted by other phenomena such as militarism: the *prima facie* strength of powerful nations with respect to weak nations. But if there are to be future international relations without war, then certainly equity is the touchstone of such a future. For the first time, the revolution of rising expectations in the Third World has been understood to entail a revolution in falling profits in the advanced nations of the First and Second Worlds. It remains to be seen whether the acquisition of basic equity demands also results in a parallel appreciation of the worth of individual liberty, or more specifically the maintenance of private initiatives.[15] Recognition that benefits for some involve costs for others is a mark of the maturation of international affairs, even though it may involve potential confrontations at a later stage of development. But as Toynbee reminded us, this capacity to absorb new challenges and creatively resolve old dilemmas is the benchmark of surviving civilizations.

Equity and energy both involve issues of basic importance, since both entail the processes of distribution and programs for redistribution. While discussions among social scientists and environmentalists alike have focused on the distribution of income and resources, the core of the problem, in the proximate future at least, is more likely to involve questions of power, since those who control the flow of capital and resources are clearly the ultimate determinants of the extent to which equity concerns are realized. This may simply be a complicated way of asserting a simple truth of our times: political considerations have largely come to prevail over purely economic factors.

An ecological approach, accurate as it may be, represents only a fraction of the story, itself tinged by global political considerations. By far, the bigger portion is the dialectic of equality and liberty. For as the state continues to underwrite programs to guarantee equity through social welfare it has been compelled to enlarge the character and scope of its own administrative apparatus. Thus, as income and wage differentials decline under the watchful and baleful eyes of the federal bureaucracy, the gap between government and citizens must widen. The very success of equity-oriented programs, involving mandated restrictions on class differentiations, serves to widen power differentiations. As authoritarian regimes have repeatedly made patently clear, the price of social welfare is political amomie for the masses. The picture that emerges is as painful as it is incontestable: libertarian goals tend to vanish as the range of federal social welfare programs expands.

For those living under Third World or internal colonial conditions of economic impoverishment, urgings toward political liberty can no longer be placed on the policy back-burner. The largely military regimes of these vast areas of Africa, Asia, and Latin America have begun to play the role of political surrogates, harnessing national economies to the achievement of social equality. The past inability of the economic system of these areas to guarantee equity through maximum participation, or the failure of the political machinery to assure liberty through mass participation has made the task of the military relatively simple. But as Third World societies become increasingly sophisticated and differentiated, the purely repressive military regime becomes less viable, serving more to disguise than to correct historical grievances and imbalances. In this sense, nations such as Venezuela, Nigeria, and Israel become extremely interesting for their exceptional features, as they attempt to deal with problems of equality and liberty in a directly civilian manner. At the same time, sophisticated forms of military-civil combinations, such as those now emerging in Brazil, also deserve close attention as possible alternatives to repressive paths of development and as options to the dire and, we hope, needless choice between equity and liberty.

For the First World and Second World no such postponement of the key issues is possible. The technical successes of industrialization under conditions of both capitalism and socialism, however limited, make the issues of political rights and participation far more volatile in the developed world than in the developing regions. It is no accident that the achievement of a large measure of social equality, far from creating a new political consensus, has only fueled concerns over the political direction of "post-industrial" societies. Just as the idea of equality once announced by the Enlightenment could not be controlled, so too the idea of liberty is increasingly central in defining the ground rules under which the dispensation of goods and services is to take place. Thus, the very success of the idea of equality—who gets what, and why—rekindles at a new level long-standing issues of liberty—who rules whom, and why.

15

Democracy and Development: Policy Perspectives

One of the great myths of our age, paid homage to by both free market and state plan advocates, is the relationship between economic systems and their political structures. Depending on the belief system at work, the illusion that democracy is a consequence of either capitalism or socialism (but not both) continues to be maintained in works too numerous to cite.[1] Yet capitalism coexists with democracy in Holland and with dictatorship in South Korea; and socialism characterizes democracy in Yugoslavia (at least in terms of worker-management), as well as dictatorship in North Korea. It is apparent that any mechanistic formula of "base" and "superstructure" cannot supplant the need for a realistic evaluation of democracy in the world at large.

The mystification of democracy in the First and Second Worlds is understandable: each wants to establish the claim that its economic system is a prerequisite to establishing a democratic order. But upon investigation, each can be seen to have idiosyncratic elements in their notion of democracy that negate the possibility of agreement about the character of democracy. As a result, analysis of democracy as a variable has given way to a theory of political democracy as a libertarian product of capitalism in the West, and a similar theory of social democracy as an egalitarian product of socialism in the East. In both worlds, democracy is held to be linked to special economic conditions which evolved in earlier historical circumstances. Such a vision of democracy is probably not relevant to present-day Third World structures; indeed, it may not even be meaningful to First and Second World structures.[2]

The subject of democracy has been widely written about. But in relation to the Third World, the tendency has been to abort serious discussion of democracy, either through a sense of embarrassment or perhaps through a sense that the subject is irrelevant with respect to the development process. For example, one widely held view is that "world systems are the only real social systems." As a consequence, it becomes meaningless to discuss Third World democracy without discussing the history of entrepreneurial civilization, the origins of the European world economy in the sixteenth century, or, even more modestly, the rise of American eminence in the late nineteenth century. In such global sweeps, policy problems are magically eliminated and hard choices nicely skirted.[3] Dealing forthrightly with democracy represents an important stage in development theory, and may facilitate the evolution of democracy itself.

Deeper thinking about the possibilities for democracy in the Third World is suggested by Guillermo O'Donnell, who suggested in a recent essay that "the issue of democracy is important not only because it contains the Achilles heel of this [bureaucratic-authoritarian] system of domination, but also because it contains a dynamic that can be the unifying element in the long-term effort to establish a society that is more nearly in accord with certain fundamental values." While he does not elucidate this normative structure, O'Donnell observes that the contradiction between the goals of democracy and the realities of daily life is

> a key to understanding the weaknesses and profound tensions of the present system of domination. It is also an indication of the immense importance of what remains implicit behind the superficial appearance of those societies who, on the one hand, are the focus of any hopes for achieving legitimacy and yet, on the other hand, are a Pandora's box that must not be tampered with.[4]

Let us examine these tensions and contradictions.

There are at least three levels at which a concept of democracy has meaning for development. First, as a simple taxonomic statement about properties of social systems that can be described as democratic. These range in type and character from personal freedom of action to public choices on behalf of the common good. Second, there is a policy context: two basic types of democratic "system"—one political, the other economic, dominate analysis. One essentially has to do with individual liberty, the other with social equity. It is fair to say that, whatever we mean by democracy, at a policy level our definitions embrace notions of extending liberty and equity. A third way of examining democracy might be called normative. Questions about democracy are related to fundamental premises of democratic civilizations and peoples: the place of obligations in a world of freedom or, conversely, the place of freedom in a world of obligations; questions about constraints upon action and the responsibility of conscience; issues relating to the role of authority in the behavior of citizens. These classical issues are well known, if not necessarily resolved.

For purposes of analysis, let us stay at the second level of discussion, concerning questions of policy as it relates to democracy. The creation of a new taxonomy is essentially formalistic; it may add more indicators or variables for consideration but shed scant light on operational or historical issues. The third, philosophic and normative level of analysis fails to address everyday possibilities. In staying at the policy level of analysis I undoubtedly betray a personal bias as well as an intellectual preference, but democracy as a question of policy rather than of language or metahistory offers the best hope for clarifying the terms of action, no less than theory. As developing regions reach beyond the broad concept of a Third World to develop more substantial and tangible manifestations of their identity, policy issues become paramount.

Notions of political and economic democracy have been juxtaposed so as to make each a repository of the First World or the Second World. Such mechanical views hold that questions of liberty and political democracy are characteristic of the Second World. Why is this so? In the First World, questions of liberty have arisen in relation to the organization of political and party life. These have involved issues about the nature of such differences as well as about the capacity of parties to mobilize the citizenry. Pluralization has turned into fragmentation as interest groups displace organized mass parties in the conduct of national political life in the advanced Western bloc. Hence, it should occasion little surprise that the relationship between parties and politics in the Third World is less than perfect. Indeed, more surprising is that multiple parties have remained a factor in a considerable majority of Third World nations.

The Second World tends toward reification. It erroneously assumes that economic democracy is uniquely characteristic of socialism and of planned societies. Stratification within the Second World is accentuated by political favoritism. There are such variations within social sectors from one nation to another that even the assumption that economic democracy is linked to Soviet socialism is subject to severe scrutiny. If factors of race, ethnicity, and sex are added to those of class, it is apparent that there is no automatic correlation between democracy and the Second World.

Many students of the Third World have been hopeful that somehow that World would synthesize political democracy and economic democracy. Unfortunately, this synthesis simply has not come about. Quite the contrary, throughout much of the Third World, the basic structure of the governing regimes remains military; the basic administrative form remains bureaucratic and authoritarian; and the basic ideology has become nationalism.

Locating sources of democracy in the Third World is no easy task. It is no more the case that democracy is consonant with political forms of rule in the First World, or economic forms of rule in the Second World, than that democracy characterizes military forms of rule in the Third World. Each World has different revolutionary origins. The capitalist system was established in the First World long before consenting or parliamentary curbs on the laissez-faire economy were undertaken. Economic changes in the Second World were wrought by parties dedicated to the establishment of some kind of socialist system; in this World, economic change followed and did not precede political revolution. In many parts of the Third World, changes came about through military intervention and military movements from below as well as from above. Economic development and political systems were largely subjected to military rule. Any automatic correlation between the Third World and the democratic system is obviously no more viable in the case of the Third World than in that of the older systems.

Political democracy is no more a bailiwick of the First World than economic democracy is a monopoly of the Second World. Democracy is neither antagonistic to nor supportive of any specific economic order. It is no more true that there exists a high level of economic democracy and a low level of political democracy in the Second World than it automatically follows that there is no search for economic democracy in the First World. The exercise of liberty and the preservation of freedom in the First World have led to a search for economic democracy as well. Democracy is a unitary phenomenon: it tends to hold together the selection of factors that make for free choice and human rights. It is a serious error to assume that democracy is parceled out as are goods and services. Bearing this in mind, the question of democracy in relation to the Third World must be considered anew, if indeed it is to be considered realistically.

Up to now there has been a strange adaptation by Third World authoritarians of the Huntington thesis that the costs of democracy are too high for the developing countries to bear. It has been assumed that high levels of development require high levels of authoritarian control or military rule. There has been a strong relationship between the military factor and the developmental impulse. But the literature also indicates that there are many cases of high levels of militarization and low development. Some forms of military rule admit of higher levels of democratic participation in the social system than theories of military domination would admit.

Today, military dominion remains the impulse behind the Third World, while there are increasing demands for democratization of the substructure. Particularly in Latin America, all sectors—Southern cone countries, Andean bloc countries, Central American countries, and Caribbean countries—have tended to call for democratic procedures at the political, economic, and social levels; at the same time each has admitted that the military factor should remain central to the developmental impulse.

Quite apart from rhetoric, the Third World shows no simple relationship between militarism and dictatorship. Militarism and democracy coexist everywhere. Democratic norms are observed in relation both to demands for personal freedom and to demands for economic justice consonant with military rule. If there can be democracy under single-party rule in the Second World or multiparty democracy in the First World, then there can be democracy under military rule in the Third World. Once we understand that democracy is an archetypical norm, covering many strata of society, then the form and the origin of revolutions in the Third World recede in importance, and the antagonistic juxtaposition of militarism and democracy can be replaced by a more accurate model.

Despite the military origins of many Third World revolutions, absolutist rule is no longer feasible precisely because military rule is now seen as a remarkably successful *stage* in the development process rather than a negation of liberty or equity. In many nations (although not all) under military rule, levels of development have risen steeply. Countries like Brazil, Peru, Mexico, or even Argentina, in a global context, make plain that as demands of development are met, as economic growth has become a social norm, concerns increasingly shift from raw aggregate growth to the equitable distribution of material goods and services. With this comes increased demands for democracy, because, whatever else it signifies, democracy has to do with distribution of goods and services and the right to express dissatisfaction over existing forms of inequity. The success of military rule in the Third World has not satisfied some social strata but has only intensified contentions for political ascendency and accentuated the "crisis of credibility" Carl Stone has written about with respect to the Caribbean region.[5]

The Third World is moving toward, rather than away from, democratic rule as a function of rapid development. As a nation develops, demands for democracy become louder and clearer. There are, to be sure, nations that do not reveal such tendencies: the two Koreas in Asia, Paraguay and Cuba in Latin America, Libya and Iran in the Middle East. Curiously, while some of these nations develop and others stagnate, the ideological commitment to one or another bloc seems to play little role or display slender linkages to democratic rule. While demands for a wider distribution of goods and services seem constant, such demands cut across specific forms of state power. The general impulse of the Third World is for widely disbursing goods and services. Less evident is the demand to disburse power and authority in a similarly holistic manner. While there is some correlation between the more advanced sectors of the Third World and democratic forms of rule, this is certainly not a uniform tendency, much less a social "law."

What undermines assumptions that democracy and development are related in the Third World is the special role performed by the military as the chief architect of social and national integration. Given this subculture of military rule, the Third World has historically had a political base that has at least made democracy possible; democracy has not been easily done away with. Throughout its history, Latin America has had a strong democratic impulse toward personal freedom. The Third World has tended to move from personal and particularistic concepts of freedom to military concepts of justice. Especially in the advanced countries of Latin America and Africa, democracy has reasserted itself as an agenda item now that development has reached take-off levels.

The Third World is faced with a set of false options. It is not simply a choice of taking from the First World its political democratic models, or from the Second World its economic distributive models, because these models have not been completely realized in either of these two worlds. The Third World must contend with militarism as the substantial base of all organization of life. The relationship of that military cluster to the practice of either economic or political democracy remains central.

Democracy is not addressed by politics or economics but has to do with health, education, welfare, and the social uses of the public domain. The military has not been able to address such social considerations in a direct form. Hence the drive for democracy in the Third World has to be judged in terms of social democracy, no less than political or economic democracy. Areas of public domain are entirely twentieth century in character. The forms of such social democracy have been worked out in the capitalist West and the socialist East with varying degrees of success. Questions of democracy can parallel questions of development. For the first time in Latin America we have the same issues that confront those who are operating in North American terms—how much democracy can a society absorb, and at what level of growth? The policy issue is no longer an absolute choice between development and democracy (or dictatorship or democracy, or militarism or democracy) but rather the peculiar stresses and strains that come from developing a democratic order in societies that imperfectly realize their economic and political goals through military means. How the exercise of democracy takes place despite the existence of militarism, which has inspired many developmental changes in the latter half of the twentieth century, is the critical question and the unique opportunity.

Democracy is a unitary concept even though we may isolate political, social, and economic factors. The exercise of personal freedom issues into ever-increasing demands for economic justice and equity. Similarly, as economic equity and economic justice are obtained, there is increasing demand for personal liberty. Certainly new demands within the Soviet Union and China for personal freedom of expression illustrate just this fact. Students of the Third World and of Latin America in particular must appreciate the extent to which the concept of democracy is volatile because of its unitary nature. It is not simply a managed policy; democracy is not simply something one can have in one area and curb in another area of public affairs.

Not only intellectuals but the military have exhibited a failure of nerve. In many parts of the Third World, neither group appreciates the degree to which small numbers of people cannot simply regulate democracy as they can levels, forms, and tempos of economic development. A political leader cannot curb personal freedom or impose a reduction of egalitarianism without having a deep impact on economic development. It is also a mistake to think that democracy must necessarily take the form of multiparty regimes on the one hand or Leninist single-party regimes on the other hand. Neither tendency has really altered events in the Third World. What has happened is that military rule and democratic thrust have become parallel. The military has become more like a policymaker, like a politician, in fact, and more like an economic manager. It has allowed a wider—far wider—area of democratic demands to surface.

The examination of democracy in the Third World cross-sectionally reveals a continuation of military forms of domination and democratic expressions of authority. It is a profound error to insist that Third World nations are garrison states with no democracy, only militarism: one must not miss the dynamics of change and the taproots of political motivation. It is also a profound error to talk about the Third World as a place without local autonomy, or without the ability to act on its own and for its own. The mistakes of the dependency model in this respect are inadvertently notorious. First this model falsifies what the Third World stands for; and, second, it falsifies the levels and the range of free expression in each nation of the Third World. The dependency model introduces a note of dangerous sentimentality in which the powerful First World (or Second World) confronts a powerless Third World. The balances struck between the major powers and the energy suppliers indicate how volatile the development patterns are, and how archaic and outdated models can become.

The Third World has raised a new question: Can democracy exist when regimes are military rather than civilian in character? For the first time we are confronted with a situation where neither multiparty or single-party economic-vanguard states are as decisive as manifestly military regimes. A syndrome evolves where military rule guarantees high levels of economic development that yield a democracy rather than a dictatorship. That is both the interesting challenge and the fact in all Latin America and much of the rest of the Third World at this time.

One of the weaknesses in the Western literature on the Third World is a tendency to cast democracy as either distributional (economic) or individual (political). This is a dangerous oversimplification. The distribution of goods and the choice of ends are both basic to development. What is "distributed" is never simply economic goods but also political power. Hence, democracy in fact has the same structural characteristics throughout three worlds of development.

As has been observed in relation to the Caribbean and Latin America, and is also true for large sections of the Middle East and Asia, the class system of the Third World itself is relatively weak. Political life is stable not simply because of egalitarian or libertarian impulses but because no specific class sector is able to come to power without opposition. A relatively weak and highly differentiated bourgeoisie confronts a relatively weak and undifferentiated proletariat in the game of politics, while large (even majority) portions of the population remain outside the political process. In the Third World economic classes remain relatively small and divided; political participation is restricted to such classes to the exclusion of the popular masses; and politics becomes an increasingly important "game" for the participants, often to the exclusion of other potential players. The character of electoral involvement reflects the fragmentary nature of democracy in the Third World.

Institutional legitimacy that derives from mass participation is relatively rare. It may be artificial when, for example, high electoral numbers are sought in meaningless exercises to sanctify acceptable officials. The ritualization of the political process in the Third World means that democracy itself becomes formalistic. Safeguards against such formalism often inhibit international growth. For example, much of the Third World consists of small nations with limited economic potential and equally limited strategic potential. And while these factors create small power resentments of big powers, they also promote democratic possibilities. Participation accords better with the sort of direct involvement possible in small nations than the sort of symbolic, massive politics of very large nations.

In discussing the possibilities for democracy in the Third World, one must not overlook tendencies that trade a weaker international role for a stronger national sovereignty. Small states have been so self-conscious about the weakness characteristic of smallness for so long that some of its more attractive features in terms of system maintenance and capacity have escaped attention. For example, a personal element can be retained in political participation; participation can remain high; economies of scale can readily be managed; and small states are exempted from at least some of the major international tensions that grip big powers.[6] In a political universe divided between big powers, small powers are often given political maneuverability they would lack in a unified world system. Democratic societies may emerge where none existed in an earlier, colonial era.

Scholars in the Caribbean region, where small nations abound, have been particularly sensitive to the possibilities of Third World democracy in a highly developmental economic climate. On the basis of a careful analysis, Edwin Jones has pointed out that administrative effectiveness and structural change can take place only if the region's small size is conceived of "first as a resource and only secondly as an obstacle." The other point, too easily overlooked, is that size and democracy correlate best when a motivated leadership is highly responsive to populations that care. Hence, small states may lead to "a developmentally relevant ideology or mobilizational system which would lead to the institutionalization of problem-solving techniques equal to the problems of a neo-colonial economy, distorted social attitudes and institutional irrelevance."[7] Even if one discounts the longings of developmentalists for the actualities of broad masses, the fact remains that a new impetus has been given to the interrelationship of democracy and development by the huge burst of the so-called mini-states, which until recently have been more patronized than perceived.

The structure of present-day world power is quite distinct from the structure of power in the era of colonial rule. Bipolar struggles characterize this period, colonial Western hegemony the earlier period. It may seem simpler to focus on the history of Western imperialism, but it is truer to events of today to appreciate that the Soviet Union has also developed a variety of imperialism. Those who urge closer economic and political links with the Soviet Union need only examine the relationship of COMECON nations to each other and their structural similarities to Western varieties of core and periphery relations. The point is that democracy is easier to maintain in a bipolar climate than in a unitary world-systems climate. The strategy of non-alignment is thus instrumental to the installation of democratic structures. This is a lesson increasingly understood by Third World nations as they chart their course between the Scylla of the West and the Charybdis of the East.

The Third World has come to appreciate that democracy is not just a system but a strategy. Like other development strategies, the democratic strategy is adopted to the degree that it does not result in total chaos at one end or total stagnation at the other. If democracy means a collapse of leadership and organization then the process of development ceases. But, just as certainly, when democracy becomes a technique to supply basic needs without providing for growth, or worse, a method for achieving equality by lopping off the heads of the opposition in a paroxysm of leveling, the overall growth needs of a society become imperiled. At that point, the same sort of risks exists in pushing the egalitarian button as exists in pushing the totalitarian button.

Third World leadership constantly monitors and evaluates the cost-benefit ratio of democratic vis-à-vis authoritarian strategies of development and does so in a context of a scarcity in managerial skills and personnel. Such leaders must be careful not to separate the developmental process from democratic impulses, since to do so will demoralize a society and eventually create sources of rebellion and reaction. Masses, after all, do not engage in insurrection in order to develop; they will, however, rebel for democratic ends. Leaders must also be careful not to confuse dialectical relationships between democracy and development with mechanical relationships between anarchy and behemoth. And here the need for evaluation research and policy studies in Third World contexts becomes paramount.

Even though democracy cannot properly be considered a "stage" of development, it is no accident that advancing Third World countries such as Nigeria, Brazil, and China are capable of movement in this direction, while more backward states are not. As Gino Germani recently pointed out, in both market and planned developing economies democracy must create full employment conditions as well as a variety of opportunities, even if this means employing people, goods, or services above and beyond the optimum required in terms of the available technology.[8] Indirect social welfare, largely determined by political concerns (which has the latent function of decreasing income redistribution), is characteristic of democratic societies. Of course, the problem is generating sufficient surpluses to "pay off" in either a patronage or client system, those pressure groups and interest groups clamoring for the further extension of egalitarianism. As Huntington has warned: the decreasing ability of democratic societies to offer such rewards in the future may jeopardize the developmental process itself.[9] Again, the degree to which democracy, like development, is a matter of policy rather than invisible forces and/or hidden hands becomes manifest when examined in a comparative context.

At the risk of suggesting an overly optimistic conclusion to a properly pessimistic decade one should take note of large-scale shifts in the political structures of major Third World powers. Comparing the beginning of the 1970s with that of the 1980s, researchers must note strong democratic tendencies in many Third World nations. A decade ago Nigeria was reeling under a recently concluded civil war, with a military regime that seemed ensconced for years to come. China was in the throes of the last stages of the Maoist gerontocracy, seemingly to be replaced by warlords intent on maintaining the fiction of development at the expense of real growth. Egypt was under the leadership of an intransigent regime intent on proving its equality only on the field of battle; Egypt did just that and then went on to extraordinary diplomatic breakthroughs abroad and major reforms at home. Brazil combined its "hard line" in politics with an economic miracle at a terrible expense to the masses but has since been thoroughly revitalizing its political system and reorienting its economic priorities.

These have been abbreviated descriptions of immensely complex regimes. Further, this has also been a period in which countries such as Chile have become enamored of development without democracy. Still, on balance, the prospects for democracy are encouraging; moreover, it is a tough-minded democracy that takes seriously the need for development and reduced stratification. In mentioning Third World powers such as Nigeria, China, Egypt, and Brazil, we can better appreciate how autonomous political systems are, in short-run terms at least, from supposed economic bases. The economic systems of these four countries are profoundly varied; levels of social stratification are also quite different; and, finally, levels of military penetration of the political process are also noticeably varied. Thus, large, pace-setting nations within the Third World have begun to turn their attention to the question of democracy and have also begun turning theory into practice.

Can we speak of democracy, as we speak of development, as having a tripartite response to the twentieth century? Certainly I have seen in my own work that economic development may be achieved through basic modes of the political process, basic modes of the economic process, and mixed strategies of both through the military process.[10] This is not the case in terms of democracy: that is to say, democracy may reflect strategic twentieth-century responses to the problem of development, but it is a unitary concept. It takes specific forms in different nations of the Third World, vis-à-vis the First and Second Worlds. However, the question of democracy cannot be viewed as being different organically or substantively in the three worlds of the developmental process. The level of democracy may differ, and policy impulses may be altered given different levels of economic development. But the Third

World has not invented or achieved unique processes of democracy. Rhetorical excesses aside, democracy represents the same kind of problem in the Third World as it does in the First and Second Worlds. It is basically an issue of paired distribution: first of goods and services, second of power and dominations. If interdependence is to replace dependence as a basic mode of policy operations, relationships based upon hierarchy, status, or charity must then be replaced by a notion of the democratic that extends the same kinds of equity in goods and powers to nations as it does to peoples.

16

The Limits of Modernity

The postwar era witnessed a near-absolute commitment to the idea of development. In this sense, the aims if not the achievements of the First and Second Worlds became the received wisdom of the Third World. There have been occasional murmurs about the shortcomings, even evils, of "Western" development; but even in a nation such as India, where under Mahatma Gandhi such fears had a certain political currency, the idea of development has taken deep root. Hence, the present sense of the limits of modernity, found in nations as diverse as Iran, Poland, and the United States, must be seen as a largely unexpected, if not entirely unwelcome, reaction by those who have endured the dogmas of development while remaining untouched by the tremors of real development.

I

Let us address the question of the limits of modernity by briefly recalling the stages in the intellectual penetration of Western doctrines of development. The first phase took place in the 1950s and corresponded with the pronouncement of an American Century; it was the idea of development as modernization. Scholars and policymakers came to view the problem of development in terms of a series of gaps: between elites sending messages and masses never receiving them; between the availability of air travel for some in contrast with those still living in a world of ox-drawn carts; between educational establishments that affected a narrow stratum of peoples in underdeveloped areas and a vast network of marginal classes of uneducated and illiterate peasants. The policy task was uniformly set in terms of an underdeveloped world in search of parity with a fully developed world.

The next phase took place in the 1960s, and was critical of this earlier point of view. It argued instead for indigenous development as an ideology; a model more appropriate to a Soviet model than an American model of development was advanced. The argument was that development is primarily a function not of modernization but of industrialization. Development should not be measured primarily in terms of creature comforts, commodity goods, or even personal health and welfare, but rather by rates and levels of industrial output and productivity. The developmental paradigm of the 1960s accepted the idea of development without qualification but denied that development had to do with modernism; instead, it argued that development had to do with industrialism. The argument in its more advanced, militant form claimed that commodity fetishism or modernization inhibited development by creating segmental stratification, an imbalance between social classes and social forces. The corollary is that industrialism overcomes commodity fetishism and creates national solidarity.

In many ways the decade of the fifties belonged to the "modernists," while the decade of the sixties belonged to the "industrialists." The developmental ideologies reflected the relative parity on a world scale of the United States and the Soviet Union in terms of acceptance of the idea that development has to do with rapid autonomy on a national level and rapid industrial buildup in economic terms. Whether it be in its American or Soviet, democratic or Stalinist forms, there was no disputing that development was good, that it was necessary, that it was in fact the name of the world game.

A new variant of the idea of development, which might be termed neo-Leninism, was introduced in the 1970s. This view of development was couched in new terminology, called "dependency theory" or the "dependency model." Its central concerns were not so much the nature of development but the sources of backwardness. These sources were held to occur anywhere but in the Third World itself. There was a global system called capitalism, which determined, through the cosmopolitan center, the character of the periphery (anywhere other than where the center was located). But even in this rather highly refined neo-Marxist perspective of the 1970s the idea of development itself was never challenged. It may have been argued that the sources of backwardness were not in domestic culture but in the international economy, not in the national capital but in overseas international capitals. But there was no disputing the virtue and the varieties of development. Commitment to the concept of social change was universal. It was expressed by modernists, industrialists, and *dependentistas* in the economic realm, as well as by a whole panoply of people in serious intellectual circles whatever their ideological bias or their political commitment. There was almost no challenge to the notion of the essential virtue of development.

A review of dominant currents of thinking and acting regarding policy over the last three decades reveals that something profoundly new is happening today. Development—indeed, the very concept of development—is under attack. The virtue of planned progress is questioned. How has this come about? This counter-revolution is related to re-creation of the Malthusian paradox; its roots have to do with the pot-of-gold theory—its realization in fact for the first time.

From the time of the conquistadores, it has been believed, sometimes covertly, that somewhere there is a pot of gold that could make the risk of economic adventure worthwhile. But when the pot of gold materialized, it was in a totally unexpected context: it happened in the Middle East. The pot of gold turned out to be "black gold"—petroleum. A whole new level of wealth came into being, created not by development but in response to a crisis in development. One discovers oil; it simply exists, although oil rigs, drilling equipment, and special maintenance are required to reach it. Oil is not a dependable consequence of hard work. And oil exists in unequal proportions in geographical terms; in this respect oil is not like food, although both are a source of energy.

The material basis for criticism of the developmental hypothesis was the spiritual feeling that God had given former "have-nots" a monopoly of a rare source of energy required by the "haves." God, a spiritual engineer, arranged for the redistribution of world power, not just energy sources. One might argue that the energy crisis of the 1970s had an impact not unlike that of a major world war: it redistributed forms of power and sources of wealth no less than energy. This realignment was accomplished without war, but in a climate of quasi-animosity between more developed and less developed nations, not because of military activity (not very much of it in any event). The East-West balance of terror remained sufficiently stable to create a vacuum, in which Middle East OPEC nations were able to negotiate their demands with impunity.

II

The sense of the limits of modernity has structural no less than historical roots. An iron law of development is that people and leaders of less developed areas observe the material consequences of development, but not necessarily the extremely difficult and complex processes of "getting there." Leaders in the Middle East in particular, by virtue of the pot of black gold, can realize the fruits of advanced development without incurring the social and material costs of older developed societies. Hence, there arises a severe imbalance between advanced commodities and technology and the infrastructure in which they can be used, an imbalance that persists over time.

A second element in the revolt against modernity is inspired by socialist societies, especially Soviet society, in which materialism is not simply a fact of empirical life but the organizing belief system of the whole society. Given the extraordinary force of religious fundamentalism within Islamic societies, and to a lesser but still noticeable degree within Catholic societies, ranging from Poland in the Soviet orbit to Brazil in the Western camp, this emphasis on materialism leads not so much to a "new man" as to a rather bizarre recycling of old capitalist man: venality, greed, and dishonesty are not so much overcome by socialist systems as they are hidden from sight and filtered through much abused humanitarian rhetoric.

A third structural source of the new fundamentalism is the

perception that economic development is a segmental activity; it is a phenomenon that stimulates growth, but it also creates wide disparities within the national culture. As a result, far from being an integrating mechanism, as it was in the industrial process, development may also function as a disintegrating machinery in pre-industrial societies. Costs of development are borne disproportionately, and, worse, rewards are also disbursed disproportionately. Arguments by defenders of Western culture that such disparities are necessary to maintain innovation, industry, and invention tend to fall on deaf ears; they confirm the belief that the developmental ideologies are intrinsically materialistic, and hence evil.

A fourth factor is the concern that any authentic egalitarian model must reject notions of change that permit, even encourage, a variety of theories concerning political vanguards and/or intellectual elites. The urge toward development is too often seen as a revolution from above. The new fundamentalism claims that such developmental ideologies widen the sense and reality of disparities between elites and masses, proletariats and peasants, urban cosmopolitans and rural locals, and so on. In short, the developmental process encourages class differentiation and poses a serious threat to national hegemony in ethnic and religious terms.

Finally, the developmental model promotes a wide array of dualisms that may negate the value of the model. It so isolates the spiritual life from the political or material life as to foster needlessly sharp and antagonistic differentiations between church and state, clerical and lay forces. In political terms, therefore, the developmental impulse isolates a series of rational measures and models that deprive social life of spiritual meaning, or a sense of teleological purpose.

Thus, the ideological groundwork for an assault on modernity—criticism of development per se—is in no small measure a result of the quixotic nature of the developmental process itself, not simply or primarily a Third World conspiracy against modernization. Development has reached such "high" levels that it has come to depend, in a kind of economic irony, upon "low" forms of energy to make the engines of change work. Development has occurred with the assistance of advanced technology, computerization, and miniaturization; but, on the other hand, development can be frustrated by Bedouin tribes controlling major sources of the world's fuel energy. And without such energy this modernization process could not occur.

The empirical paradox took on a metaphysical character. Ultimately there was a crisis about what really counted. Providence, not socialism, was said to account for the monopoly of fuel energy in the hands of have-nots, energy without which the entire developmental process would come to a crashing halt. Such monopoly, furthermore, did not make it necessary for such nations to employ the equivalent of bows and arrows against atomic weapons. The disparity between military power and economic domination itself became a proof of Divine support for the anti-modernist vision.

A serious consequence of the theory of petroleum as a pot of gold is that it has limited commodity characteristics. An agricultural or commercial system creates a business civilization, but oil creates only revenues. As a result, an economy based on oil can afford the luxury of anti-developmentalism as an ideology, since it can import from abroad the sorts of managerial and organizational skills Western societies must generate from within. Beyond that, the pot-of-gold economy precludes the sort of developmental regimes found earlier among the Japanese Shogunate and the German Hohenzollerns. The maintenance of a pot-of-gold economy permits sheikdoms lacking mass support to retain power. But traditional tribal forms of rule can maintain their varieties of Islamic fundamentalism only for that period of time in which the petroleum pot can be replenished. The structural fragility and temporal limits of an economy based on petroleum cast a long shadow over the revolt against modernity in these cultures.

The basis of the ideology of the 1980s is a sense of the limits of modernity and development as such. Once a revolution in the distribution of goods was achieved, and once an absolute paralysis in the fueling mechanism or triggering mechanism of the advanced sector could be produced, if need be, a shift in the theory and practice of development became inevitable. For the first time the modern world is being attacked.[1]

The decisive point is not the role of religion in the organization of state power but, rather, the uses of religious symbolism to confound the notion of development. In earlier epochs, whether in Italy, France, or England, religious fervor was used to organize state power to enhance the developmental process. What uniquely characterizes the current era is *the organization of religious forces either through or against the mechanism of state power* to frustrate the developmental process. This is an important difference. Religious fundamentalism has come to characterize the present period and current mood.

Religious fundamentalism is not only a Middle Eastern phenomenon. Poland is interesting in this regard. Few photographs of the workers' strikes of 1980 failed to show a group of workers displaying a crucifix or in the background a picture of the current Pope. The organizing premise, the ideological formation, was not only more money and less work—ordinary working class demands—but a strong ideological commitment to Christianity as an answer to the industrializing theories of Marxism. In both Iran and Poland, at some level the use of religion, or the use of traditionalism if one prefers, is aimed directly at the heart of the concept of development.

What shibboleths of development went unquestioned in the past? First, the need to sacrifice for the next generation and in the process to display a primitive accumulation of wealth. Demands were confounded. On the one hand came the call to sacrifice, on the other the assertion that material goods are not worth the sacrifice, that they are shabby, that they have no ultimate worth. A second shibboleth is that a society is defined by the character of its developmental process. The counter-argument is that society is not defined by the developmental process, but that the developmental process tends to wash away the unique characteristics of each society or each civilization. A conflict between traditionalism and modernity is emerging that threatens the developmental process itself. This is much more than a theological phenomenon. The zero-growth movement, the limits-to-growth movement, the idea of zero growth as a positive good—all assert that there are abstract values, goods, and services quite beyond those resulting from concrete development.

What potential do these movements have for frustrating further industrial innovation? How seriously should they be taken? What is the relationship between developmentally oriented regimes or developmentally oriented sectors within developing societies and the traditional elements or theocratic elements within these societies? In certain parts of Latin America a tremendous struggle may be taking shape between a military sector that is oriented toward development and a religious sector that is oriented toward traditional values. The issue is akin to the argument concerning indigenous folklore, about the relationship between the international environment and its folkloristic roots. The entire direction of the twentieth century is in question. The measurement of society or civilization by a gross national product, by levels of industrial output, or by levels of consumptive activity has come under tremendous criticism. At some level, the consequences will be a redefinition of what constitutes value as such; what constitutes the relationship of base and superstructure as such; and what constitutes the relationship of state power and religious power as a mobilizing force.

The ability of this new anti-modernization movement to survive is not likely to rest primarily on the structure of economic productivity but more decisively on the character of nationalist impulse. For the weakening of the modernizing sector also implies a lessening of the militarizing sector; and that means the essential pillar of power upon which modern nation-states rest is much more subject to external threat and military adventure. In Iran, the main danger to the rule of the religious mullahs came not from the modernizing forces, which were overthrown, but from rival regimes in the area such as Iraq or the Kurd separatists, who saw the opportunity to chip away at what has been a major force in the area. Whether religious fundamentalism can withstand an impairment of nationalism is an essential touchstone both of the depth of feeling involved in this revolt against modernity and the survival capacities of the counter-revolutionary regime which seizes power.

III

What is the impact on the United States of this revolt against modernity? This question concerns not only the United States but in part the Soviet Union as well. The postwar environment was shaped by the struggle between the United States and the Soviet Union, involving competing concepts of development between the

Keynesian concept in the West of modernization based upon consumer satisfaction and the Marxist concept of industrialization based upon principles of national security and self-sufficiency. The entire postwar dialogue has been about which type of development should be implemented. No question has been raised about whether development is good per se, only about which form of development is admissible.

The rise of a special Third World segment, high in energy resources and otherwise low in developmental skills, has had a very profound impact on American society. For the first sustained period of time in its more than two-hundred-year history overseas reality is affecting everyday life in America. Unlike Europe, the United States has never been affected directly by war or famine or any other ravage. Even World Wars I and II, the Korean War, and the Vietnam War were essentially overseas events. People lost loved ones, and war politicized the domestic climate of opinion, but essentially these wars had no mass impact on American life. The material life of American society remained largely unaffected by events that took place overseas.

This changed dramatically with the 1973 oil embargo; in the 1978–79 oil crises, the ordinary American felt that foreign affairs were real, more so than they had been at any previous time. The relationship between what went on abroad and what was going on at home became direct—in its reality, in its impact, and in its consequences. What began as a strategy and a tactic of Middle East regimes and the OPEC cartel became a world historic event in American life. For the first time Americans understood that they were not self-sufficient. Even the Vietnam War, which challenged the concept of American invincibility, had not shaken that feeling of self-sufficiency.

Every Soviet leader—Lenin, Stalin, Khrushchev, Brezhnev—has aspired to the United States' level of self-sufficiency. The idea of self-sufficiency is the ultimate metaphysical payoff of the developmental process. Why develop? Why accelerate development? The answer is always that self-sufficiency is desirable. The nation will no longer be beholden to, no longer be dependent upon, other nations. The events within Middle Eastern societies like Iran or Saudi Arabia or other OPEC nations provide a framework within

American life for challenging the developmental thesis in its essential form—the doctrine of national self-sufficiency.

At first the challenges to modernity took a secular form in the West, which might be identified as environmentalism, which asserted that the environment had to be preserved, even at the cost of economic growth. Industrial waste and industrial diseases were too high a risk factor; nature had its own value. Certainly, the early phases of the environmental movement were relatively innocuous. There was nothing particularly theocratic about it; the environmental impulse was, if anything, naturalistic. Environmentalism embodied a belief that somehow the relationship between environment and industrialism had to be dealt with in an entirely new way. Admittedly there were special problems involved: labor reallocation, increased fuel costs, increased costs of goods. The presumption was that these problems, as well as reduced industrial development and modernization, were worth it because nature had its own value.

The second round of the challenge to modernity moved beyond the environmental—beyond the purely secular. The rise of religious cults and of fundamentalist movements meant that the rejection of modernism had extended to a re-evaluation of the notion of what constitutes a good world. Limited growth was quickly translated into zero growth. Spiritual values were emphasized over material values. Millions of people within American society began to participate in religious fundamentalism. The origins of these sentiments were not so much anti-developmentalism as a spiritual answer to the externally imposed limits to further material growth. No growth became better than slow growth because it was easier to assert as a metaphysical first principle. The naturalistic impulses of the environmental movement of the seventies were transformed into an aggressive spiritual revolt against modern values, not infrequently involving the same fears of advanced technology and the same appeals to puritan values of self-discipline.

A third level was a kind of right-wing impulse toward antimodernism, akin to isolation from the world at large and from America's problems in particular. What has evolved within the United States is unadorned anti-developmentalism akin to chiliastic visions in the Middle East, and no less demanding in ideological terms. What remains to be examined is how the new fundamentalism relates to engineering, science, and research. How does it connect with advanced forms of development within American culture and society?

At the very time a new kind of fundamentalism arises, new technological developments have occurred in data-base computation and in activities related to electronics, polymers, and biogenics. The world of science and information is accelerating at the same time that religious fundamentalism is expanding. Events in the West seem to parallel what is taking place in the Third World generally, and within the Middle East specifically. This leads to a two-track system, a religious track and a scientific track; as a result, development is increasingly linked to elite rather than to mass requirements.

There is a gigantic cultural struggle under way: the culture of development versus the culture of the spiritual; the culture of the traditional versus the culture of the modern. We are experiencing a new version of the nineteenth-century struggle between science and religion.[2] It does not so much involve general ideological formations as it does the consequences of this ideology for development of power in the world. What are the consequences of this kind of traditionalism for Eastern Europe? For socialism and communism? For Western concepts of democracy?

One can only surmise what the consequences may be. There may be a kind of worldwide isolationism. Within the United States, we may experience a period in which these new movements—whether they be pentacostal, environmental, or libertarian—create dogged and determined neo-isolationism, while the nation protects itself through advanced technology and advanced systems design. The divisions between the scientific and the spiritual allow American society to develop what may be called capital-intensive militarization through advanced forms of scientific endeavor. On the other hand, they also encourage mass participation in theological and neo-isolationist movements, in reaction to this increasing sophistication. Among the masses, non-participation and non-commitment have become normative. Even under the most extreme provocation, U.S. political leadership, whether Democratic or Republican, must be cautious about extending its claims. The polarization of scientific and religious culture and the assault on development severely limit any claims the leadership can make on the masses.

A potential outcome of fundamentalism within the United States is authoritarian domestic politics. The character of American society is shaped by mass forces. Fundamentalist frameworks have to be expressed and could take the form of a protest not only against modernism but against presumed excesses of sexual liberation or excess personal freedom in general. At some level the new fundamentalism may challenge the pluralistic value base of American society. American society may not move in an authoritarian direction, but certainly it will if there is an economic slowdown coupled with a steady rate of high unemployment, and as a result this produces higher demands on the system. Within a stern moralistic value system, obligations to the system are emphasized over rights within the system. The consequence may well be an exaggerated turn to the Right. What limits such fundamentalism is the huge shift from rural to urban patterns, and the general secularization of culture in personal habits and decisions. Still, there is no incongruity between an authoritarian-bureaucratic orientation domestically and a neo-isolationist foreign policy; they may work very well in tandem. Shrinking power is increasingly focused on domestic rather than foreign policy.[3]

Anti-modernity, anti-developmentalism, and anti-industrialism are phenomena that could lead to an American society increasingly insular in character, isolated from world patterns, and subjected to the same internal pressures that have led to "Finlandization" in Europe. A revolt against a complex present leads to policy-making unilateralism, disregard of the "other" side, and the sentimental view that all real problems are negotiable between "men of good will," whatever the nature of the social system.

The secular traditions, including French Enlightenment, German Enlightenment and romanticism, and American modernism, have all experienced a peculiar crisis. Requirements for personal achievement have become increasingly complicated. There is a systems overload within modern culture (and here I include the Soviet Union as well) because each generation must confront so much information, digest so much material, absorb so much innovation, and store so much scientific knowledge. The overload tends to produce an opposite result: namely, an impulse to disgorge or empty out. The developmental model came upon hard times not so much because of its negative consequences but through its positive results. Managing these developmental processes proved so complex that they became problems instead of advantages.

Take one small example. The phonograph machine is an entirely twentieth-century artifact. It began as a simple, hand-driven turntable, coupled with a record made of shellac which at best gave modest sound reproduction. Digital recordings are now so precise that recordings only three years old have become obsolete. The sophistication in equipment makes what we generated a decade ago appear primitive in comparison. Word processing equipment requires almost professional-level knowledge to assemble the equipment necessary to achieve the desired results. Similarly, in the developmental process, an enormous case of information overload occurs, resulting in an incapacity to enjoy the results of the developmental process or a much higher level of coping, as with a vast network of home computerization.

Anti-developmentalism becomes a critique of complexity as such. It is not only an attack on the notion of gross national product but a response to the difficulty of absorbing the language of mathematics and computer science or the methods of research and theories of evidence. The world of science and technology has become exceedingly difficult. The fact that significant numbers of people are unable to absorb large chunks of information has created a special problem within American culture, especially since it stands as the most advanced "modern" society. A rebellion thus arises not just against the developmental, but against complexity and difficulty. The failure to establish a firm set of answers produces frustration. A world in quest of certainty is denied answers. Instead, the developmental paradigm presents problems to be researched and policies to be evaluated. Teleological resolutions of life arise as a challenge to the new technology necessary to cope with the continuation of the developmental impulse.

It is extremely difficult to develop a scientific or technological style that encourages a developmental process within a society. In American society, no less than elsewhere, development has always been an elitist concept, a slogan put forward in the name of the people. But, basically, development remains a vision held by those at the top who have a sense of the national conscience and the national consensus. The cult of the traditional is a rebellion against such elitism. It is a demand for simplification, for a world in which answers are known. The current wave of fundamentalism in American life rests on ideology based on solutions—truth through Providence. In this sense, the fundamentalism of the Middle East may have a direct relationship to similar events in American society. The assault on modernity within American life and elsewhere should be taken with absolute seriousness. It affects the character of individual life, community values, and ultimately the nature of state power.

One Step Forward, Two Steps Backward

There remain several unanswered questions. Is this new moralism and religious revivalism simply a cyclical event in the United States, as shortlived as the "hang loose" ethic of the 1960s or the "environmentalism" of the 1970s, or will it be more durable? Will the isolationism that derives from such a moralism be any different from similar tendencies following World War I, with its Kellogg-Briand Pact outlawing war, U.S. rejection of League of Nations membership, and general dismantling of armed forces? These considerations pertain to temporal, cyclical aspects of what has herein been discussed in spatial and global terms.

While it is clear that there is a loose connection between the generational characteristics described above, it would be dangerous to reduce the new fundamentalism to such a common ancestry. What we are witnessing is not simply a challenge to modernity but an assault upon complexity, especially against scientific findings and formulas. The religious mood of exaltation and fervor, even if it remains a statistical minority, is a relatively painless way to knowledge. In place of many books is The Good Book; in place of relativism is moral certainty; in place of a series of questions begetting more questions are a series of answers stimulated by the rhetoric of certainty.[4]

There has not been a serious, full-scale appraisal of the concomitant to the end of ideology: namely, the exhaustion of mythology. The scientific ethic—with its emphasis on experience and verification, information that is subject to testing and grounded in methods, data presented in a highly mathematical form—inevitably leads to a new elitism, based upon policy-oriented varieties of professionalism. There is a mushrooming of agencies based upon evaluation, measurement, policy-making—all of which debrief, debunk, and move people away from common inherited shibboleths. The new chiliastic religions and cults provide a wide variety of answers in a world of doubt, certainty in a world of uncertainty, and belief in a world of competing facts. The structure of scientific revolutions, no less than the structure of foreign policy, provide the motor force of this moral counter-revolution against economic development. The force of numbers, of masses, is potent. The power of mobilization has been crucial in the postwar climate from Gandhi to Khomeini. Such a force prevents limits to military might from being operative in the face of mass movements from below. The new moralism does not shy away from politics. Rather, it denies that politics is a complex game played by the rulers alone. Hence, the new fundamentalism perceives itself as supporting the restoration of traditional values, including the value that everyone counts in this new massive order of things.

With respect to the novelty, or lack thereof, of this new fundamentalism—whether it is simply a cyclical reaction or a programmed restoration of the isolationist tendencies of earlier decades—requires both analytic clarity and historical specificity. The old isolationism was essentially a political maneuver, a legislative and presidential tendency reflecting strong fears on the part of Americans of becoming linked to European embroglios: overseas adventures that would once again require shedding American lives to "bail out" European varieties of nationalism. Whether this scenario was accurate or not was less important than the fact that it was widely believed. The old isolationism was from the outset a political tactic reflecting cultural insularity. The American economy of the 1920s continued to expand and flourish. Indeed, perhaps no decade before or since saw such a startling expansion of American economic might and muscle. Hence, while the 1920s witnessed isolationist politics, it also revealed internationalistic economics. Political isolationism and economic imperialism confronted each other in uneasy truce. That a certain perceptual and linguistic confusion ensued, with a misperception of the relationship between a politics of noninvolvement coupled with an economics of total dominion, doubtless makes the past seem like the present. But it would be a mistake to see history in mechanistic terms. The isolationism of the past was built upon a series of strategic considerations of growth; the isolationism of the present is built upon less precious metal: how to limit growth without inviting immediate collapse; or more prosaically, how to save the economic system without an absolute reliance upon the political structure.

The present isolationism involves economic considerations: high tariff walls against imports, protection for native manufactured goods, balanced export-import relations. But these aims reflect a growth in realization that economic control is in the hands of multinational corporations that do not move in lock step with American wishes any more than Dutch multinationals take their marching orders from The Hague.[5] The globilization of capitalism comes at the expense and not necessarily with the aid of American capitalism. Thus, the new isolation often reflects less policy decisions than structural tendencies toward the realignment of the international economy. This realignment of the multinationals, coupled with the monopolization of vital energy resources by the OPEC powers, has meant that the new isolationism, unlike the old isolationism, is not something that can be turned off or on at will. The new isolationism is a reflection of a weakening, deteriorating global position. The old isolationism was a reflection of a certain quiet arrogance, a clear belief in the superiority of the United States way of life—a system capable of delivering the goods; it was the luxury of a secure imperial force. The new isolationism, by contrast, is a necessity, a consequence of a revolution of falling expectations, rather than the confidence of world supremacy. America searches for a place within the colonial cosmos, and abandons its pretentions to be at the head.

Are there measures at the policy level that could be implemented to turn this situation around? These policies must become essentially pedagogic: the creation of a scientific culture capable of fending off and withstanding the new religious revivalism. In this, the rising popular interest in intergalactic travel, space stations, planetary missions, may been seen to function as a counteractive myth, reinstituting the idea of the centrality of development into the myth of external progress. There is also a need for parallel, matching social myths. Here the problem is simply that the overwhelming myth of socialism so long dominant in the century after 1848 has broken down into fragments that cannot be reassembled. The collapse of the socialist alternative, certainly in its Soviet form at least, represents the failure of the last great secular myth. The return to traditionalism is in no small measure a function of this degeneration within the advanced societies. The industrial ideologies of both the First and Second Worlds have partially failed. Under the circumstances, new myths, of a more millenarian sort, have become widespread throughout the Third World.

The depth or breadth of this new fundamentalism, whatever varieties it takes, will be determined by the character and consistency of the response from the scientific and secular communities. The organizing values of social myths should not be minimized. The question is rather whether such myths have sustaining power. The struggle between developmental, modernizing, scientific ideologies and traditional, moralizing, theological ideologies has entered an advanced stage. We are not simply involved in a world of intellectual continuities and discontinuities but where there are quite practical and painful policy choices about whether to go forward, stand still, or move backward. The decisions taken will determine world socio-political priorities for at least the balance of the century.

Notes

1: Realignments of Three Worlds in Development

1. Wassily Leontief, et al., *The Future of the World Economy*. New York: Oxford University Press, 1977.
2. Irving Louis Horowitz, *Equity, Income, and Policy: Comparative Studies in Three Worlds of Development*. New York and London: Praeger/Holt, Rinehart and Winston, 1977, pp. 5–21.
3. Paul Wohl, "Soviet Bloc Disappoints Third World," *Christian Science Monitor*, June 9, 1977.
4. Robert S. McNamara, *Accelerating Population Stabilization through Social and Economic Progress*. Washington, D.C.: Overseas Development Council, 1977, pp. 45–52.
5. Carlos Massad, "The Revolt of the Bankers in the International Economy: A World without a Monetary System," *CEPAL Review*. Santiago, Chile: United Nations Economic Commission for Latin America, 1976, pp. 3–82.
6. Philip E. Converse, "Societal Growth and the Quality of Life," in Amos H. Hawley (ed.), *Societal Growth: Processes and Implications*. New York: Free Press/Macmillan, 1979, pp. 249–64.

2: Traditionalism, Modernization, and Industrialization

1. Daniel Lerner, *The Passing of Traditional Society*. Glencoe, IL: The Free Press, 1958; Wilber Schramm (ed.), *Mass Communications*. Urbana: University of Illinois Press, 1949; and Bernard Berelson, "Communications and Public Opinion," in Bernard Berelson and Morris Janowitz (eds.), *Reader in Public Opinion and Communication*

(enlarged edition). New York: Free Press of Glencoe, 1953, pp. 448–62.

2. Peter L. Berger, *The Sacred Canopy: Elements of a Sociological Theory of Religion.* Garden City, NY: Doubleday & Co., 1967; see also, Peter L. Berger, Brigitte Berger, and Hansfried Kellner, *The Homeless Mind: Modernization and Consciousness.* New York: Random House, 1973.
3. Fernando Henrique Cardoso, *Capitalismo e Escravidao no Brasil Meridional.* São Paulo: Difusão Europea do Livro, 1962; see also Gideon Sjoberg, "Traditional and Transitional Societies: Folk and 'Feudal' Societies," *American Journal of Sociology*, Vol. 58, No. 2 (1952), pp. 231–39.
4. Myron Weiner (ed.), *Modernization: The Dynamics of Growth.* New York and London: Basic Books, 1966; and David E. Apter, *The Politics of Modernization.* Chicago and London: University of Chicago Press, 1965.
5. John J. Johnson, *The Military and Society in Latin America.* Stanford: Stanford University Press, 1964; and *Political Change in Latin America: The Emergence of the Middle Sectors.* Stanford: Stanford University Press, 1958.
6. Gino Germani (ed.), *Modernization, Urbanization and the Urban Crisis.* New Brunswick, NJ: Transaction Books, 1973; and *Marginality.* New Brunswick, NJ: Transaction Boooks, 1979.
7. Irving Louis Horowitz, *Three Worlds of Development: The Theory and Practice of International Stratification.* New York and London: Oxford University Press, 1972.
8. Gunnar Myrdal, *Asian Drama: An Inquiry into the Poverty of Nations* (in three volumes). New York: The Twentieth Century Fund, 1968.
9. Paul A. Baran, *The Political Economy of Growth.* New York: Monthly Review Press, 1957; and with Paul M. Sweezy, *Monopoly Capital: An Essay on the American Economic and Social Order.* New York: Monthly Review Press, 1966.
10. Otto Nathan, *The Nazi Economic System: Germany's Mobilization for War.* Durham, NC: Duke University Press, 1944.
11. Kenneth E. Boulding, *Peace and the War Industry.* Chicago: Aldine Publishing Company, 1970; and *Conflict and Defense: A General Theory.* New York: Harper & Row, 1962.
12. John Kenneth Galbraith, *The New Industrial State.* Boston: Houghton Mifflin, 1967; and *American Capitalism: The Concept of Countervailing Power* (revised edition). Boston: Houghton Mifflin, 1956.
13. Baldwin Ranson, "The Limits to Growth: Is Ayres's Position Unwarranted?" *Journal of Economic Issues.* Vol. XIII, No. 3 (September 1979), esp. pp. 656–63.
14. Donald J. Lecraw, "Choice of Technology in Low-Wage Countries," *The Quarterly Journal of Economics.* Vol. XCIII, No. 4 (November 1979), pp. 631–34.

15. Robert M. Bernardo, *The Theory of Moral Incentives in Cuba.* University, AL: The University of Alabama Press, 1971.
16. S. N. Eisenstadt, "Breakdowns of Modernization," in Jason L. Finkle and Richard W. Gables (eds.), *Political Development and Social Change.* New York: John Wiley & Sons, 1966, pp. 573–91; and his more complete work in this area, *The Political Systems of Empires.* New York and London: Free Press of Glencoe/Macmillan, 1963.
17. Walt W. Rostow, *The Stages of Economic Growth: A Non-Communist Manifesto.* Cambridge: Cambridge University Press, 1960; and, more recently, his *Politics and the Stages of Growth.* Cambridge: Cambridge University Press, 1971.
18. Seymour Martin Lipset, "Observations on Economic Equality and Social Class," *Equity, Income, and Policy: Comparative Studies in Three Worlds of Development.* New York and London: Praeger/Holt, Rinehart & Winston, 1977, pp. 278–86.

3: Social Planning and Economic Systems

1. Amitai Etzioni, "Toward a Theory of Societal Guidance," in Sarajane Heidt and Amitai Etzioni (eds.), *Societal Guidance: A New Approach to Social Problems.* New York: Thomas Crowell, 1969, p. 371.
2. A. A. Zvorykin, "The Development of the Productive Forces in the Soviet Union," in G. V. Osipov (ed.), *Industry and Labour in the USSR.* London: Tavistock Publishers, 1966, pp. 16–17.
3. Friedrich A. von Hayek, *The Road to Serfdom.* London: Routledge and Kegan Paul, 1944.
4. Karl R. Popper, *The Open Society and Its Enemies.* London: Routledge and Kegan Paul, 1945.
5. Barbara Wootton, *Freedom under Planning.* Chapel Hill, NC: The University of North Carolina Press, 1945.
6. Milton R. Friedman, *Capitalism and Freedom.* Chicago: University of Chicago Press, 1962.
7. Robert Nisbet, "The Public Interest and Community Participation: Commentary," *Journal of the American Institute of Planners,* Vol. 39, No. 1 (January 1973), pp. 3, 8–9.
8. Bernard J. Frieden and Robert Morris (eds.), *Urban Planning and Social Policy.* New York: Basic Books, 1968.
9. Albert Lepawsky, "The Planning Apparatus: A Vignette of the New Deal," *Journal of the American Institute of Planners,* Vol. 42, No. 1 (January 1976), pp. 16–32.
10. Robert S. Bolane, "Community Decision Behavior: The Culture of Planning," *Journal of the American Institute of Planners,* Vol. 35, No. 5 (September 1969), pp. 308–14.
11. Barclay M. Hudson, Martin Wachs, and Joseph L Schofer, "Local Impact Evaluation in Design of Large-Scale Urban Systems," *Journal of American Institute of Planners,* Vol. 40, No. 4 (July 1974), pp. 255–65.

12. Alan S. Kravitz, "Mandarinism: Planning as Handmaiden to Conservative Politics," in Thad L. Beyle and George T. Lathrop (eds.), *Planning and Politics*. New York: Odyssey Press, 1970, pp. 240–67.
13. Leonard C. Moffitt, "Values Implications for Public Planning: Some Thoughts and Questions," *Journal of the American Institute of Planners*, Vol. 41, No. 6 (November 1975), pp. 397–405.
14. Kenneth D. Benne, Warren G. Bennis, and Robert Chin, "Planned Change in America," *The Planning of Change*. New York: Holt, Rinehart and Winston, 1969, pp. 28–32; Kenneth D. Benne, Warren G. Bennis, and Robert Chin, (eds.)
15. Henry Fagin, "Advancing the 'State of the Art,'" in Ernest Erber (ed), *Urban Planning in Transition*. New York: Grossman Publishers, 1970, p. 133.
16. Harvey S. Perloff, "Common Goals and the Linking of Physical and Social Planning," in Bernard J. Frieden and Robert Morris (eds.), *Urban Planning and Social Policy*, pp. 346–59.
17. Herbert J. Gans, "City Planning in America: A Sociological Analysis," *People and Plans: Essays on Urban Problems and Solutions*. New York: Basic Books, 1968, p. 25.
18. Magoroh Maruyama, "Human Futuristics and Urban Planning," *Journal of the American Institute of Planners*, Vol. 39, No. 5 (September 1973), pp. 346–57.
19. William L. C. Wheaton and Margaret F. Wheaton, "Identifying the Public Interest: Values and Goals," *Urban Planning in Transition*, pp. 152–64.
20. Francine F. Rabinovitz, *City Politics and Planning*. New York: Atherton Press, 1969, pp. 79–117.
21. Murray L. Weidenbaum, "Government versus Business Planning," *Monthly Labor Review*, Vol. 99, No. 5 (May 1976), p. 35.
22. Donald A. Schon, Nancy S. Cremer, Paul Osterman, and Charles Perry, "Planners in Transition: Report on a Survey of Alumni of M.I.T.'s Department of Urban Studies, 1960–71," *Journal of the American Institute of Planners*, Vol. 42, No. 2 (April 1976), pp. 193–202.
23. Thad L. Beyle, Sureva Seligson, and Deil S. Wright, "New Directions in State Planning," in Thad L. Beyle and George T. Lathrop (eds.), *Planning and Politics: Uneasy Partnership*. New York: Odyssey Press, 1970, pp. 14–34.
24. John Friedmann and Barclay Hudson, "Knowledge and Action: A Guide to Planning Theory," *Journal of the American Institute of Planners*, Vol. 40, No. 1 (January 1974), pp. 2–16.
25. Martin H. Krieger, "Some New Directions for Planning Theories," *Journal of the American Institute of Planners*, Vol. 40, No. 3 (May 1974), pp. 156–63.
26. John Friedmann, "The Public Interest and Community Participation: Toward a Reconstruction of Public Philosophy," *Journal of the*

American Institute of Planners, Vol. 39, No. 1 (January 1973), pp. 2–7.
27. Robert Nisbet, *Twilight of Authority*. New York: Oxford University Press, 1975, pp. 9–10.
28. Herbert J. Gans, "The Public Interest and Community Participation: Commentary," *Journal of the American Institute of Planners*, Vol. 39, No. 1 (January 1973), pp. 3, 10–12.
29. Frances Fox Piven, "Planning and Class Interests," *Journal of the American Institute of Planners*, Vol. 41, No. 5 (September 1975), pp. 308–10.
30. Darwin G. Stuart, *Systematic Urban Planning*. New York: Praeger, 1976, pp. 284–92.
31. Seymour J. Mandelbaum, "On Not Doing One's Best: The Uses and Problems of Experimentation in Planning," *Journal of the American Institute of Planners*, Vol. 41, No. 3 (May 1975), pp. 184–90.
32. Herbert J. Gans, "The Need for Planners Trained in Policy Formation," *Urban Planning in Transition*, p. 242.
33. Irwin Abrams and Richard Francaviglia, "Urban Planning in Poland Today," *Journal of the American Institute of Planners*, Vol. 41, No. 1 (July 1975), pp. 258–69.
34. Aaron Wildavsky, *Budgeting: A Comparative Theory of Budgetary Processes*. Boston: Little Brown and Company, 1975, pp. 258–59.
35. Yevsei G. Lieberman, "Waiting for a Bath—and Just Waiting," in Myron L. Joseph, Norton C. Seeber, and George L. Bach (eds.), *Economic Analysis and Policy*. Englewood Cliffs: Prentice-Hall, 1971, pp. 596–97.
36. G. I. Popov, "Speech on the Soviet State and Budget," in David Lane (ed.), *Politics and Society in the USSR*. New York: Random House, 1971, pp. 170–71.
37. N. N. Kachalov, "Speech on the Soviet State and Budget," *Politics and Society in the USSR*, pp. 171–72.
38. Jan Marczewski, *Crisis in Socialist Planning; Eastern Europe and the USSR*. New York: Praeger, 1974, pp. 227–43.
39. Joseph S. Berliner, *The Innovation Decision in Soviet Industry*. Cambridge, MA: MIT Press, 1976, p. 258.
40. Herbert J. Gans, "City Planning in America: A Sociological Analysis," *People and Plans: Essays on Urban Problems and Solutions*. New York: Basic Books, 1968, p. 25.
41. Murray L. Weidenbaum, "Reforming Government Regulation of Business," *Imprimus*, Vol. 5, No. 6 (June 1976), pp. 1–6; see also his work, *Matching Needs and Resources: Reforming the Federal Budget*. Washington, DC: American Enterprise Institute, 1973.
42. David Berry and Gene Steiker, "The Concept of Justice in Regional Planning: Justice as Fairness," *Journal of the American Institute of Planners*, Vol. 40, No. 6 (November 1974), pp. 414–21.
43. Barrie J. Greenbie, "Social Territory, Community Health and Urban

Planning," *Journal of the American Institute of Planners*, Vol. 40, No. 2 (March 1974), pp. 75–82.

44. William Gorham and Nathan Glazer, *The Urban Predicament*. Washington, DC: The Urban Institute, 1976.
45. Karl Mannheim, *Freedom, Power and Democratic Planning*. New York: Oxford University Press, 1950; see also Peter Marcuse, "Professional Ethics and Beyond: Values in Planning," *Journal of the American Institute of Planners*, Vol. 42, No. 3 (July 1976), pp. 274–82.

4: Zero-Growth Economics and Egalitarian Politics

1. Lester Thurow, *The Zero-Sum Society: Distribution and the Possibilities for Economic Change*. New York: Basic Books, 1980, pp. 11–12.
2. Gunnar Myrdal, "On the Equality Issue of World Development," *World Issues*, Vol. 1, No. 1 (Oct.-Nov. 1976), pp. 3–5.
3. Jay W. Forrester, *World Dynamics* (second edition) Cambridge, MA: Wright-Allen Press, distributed by M.I.T. Press, 1973, pp. 129–32.
4. S. M. Miller and John Hoops, "Work," *New Society*, Vol 36, Whole No. 714 (10 June 1976), pp. 582–83.
5. *Rural America*, "Call to the Second National Conference on Rural America," Vol. 1, No. 12 (October 1976), p. 1.
6. Center for the American Woman and Politics, *Report of CAWP: 1974–1975*. New Brunswick, NJ: Eagleton Institute of Politics, Rutgers University, 1975, pp. 6–7.
7. Hans H. Landsberg, *Energy and the Social Sciences: An Examination of Research Needs*. Washington, DC: Resources for the Future, Inc., 1974; and Mancur Olson and Hans H. Landsberg (eds.), *The No-Growth Society*. New York: Norton, 1974.
8. Leon H. Keyserling, *Full Employment without Inflation*. Washington, DC: Conference on Economic Progress, 1975, p. 9.
9. See the exchange on this subject between Kingsley Davis and Wilbert E. Moore, "Some Principles of Stratification," *American Sociological Review*, Vol. 18, No. 4 (1953), pp. 448–57.
10. Martin Rein, "Stratification and Social Policy," *Social Science and Public Policy*. New York: Harmondsworth, 1976, pp. 171–209.
11. Fred Hirsch, *Social Limits to Growth*. Cambridge, MA: Harvard University Press (A Twentieth Century Fund Study), 1976, pp. 11–12.
12. Herbert J. Gans, "The New Egalitarianism," in Lee Rainwater (ed.), *Social Problems and Public Policy: Inequality and Justice*. Chicago: Aldine Publishing Company, 1974, pp. 20–21.
13. Alice Rossi, "Sex Equality: The Beginnings of an Ideology," in ibid., pp. 260–68.
14. S. M. Miller and Pamela Roby, *The Future of Inequality*. New York: Basic Books, 1970, pp. 121–23.

15. Samuel P. Huntington, "The Democratic Distemper," *The Public Interest,* Whole No. 41 (Fall 1975), pp. 29–30.
16. Kingsley Davis and Wilbert E. Moore, "Some Principles of Stratification," *American Sociological Review*, Vol. 10, No. 2 (1945), pp. 242–49.
17. Christopher Jencks, et al., *Inequality: A Reassessment of the Effect of Family and Schooling in America*. New York: Harper and Row, 1972, pp. 9–10.
18. John H. Goldthorpe, "Social Inequality and Social Integration," in Lee Rainwater (ed.), *Social Problems and Public Policy: Inequality and Justice*, p. 136.
19. H. Pant Chalfant, "Correlates of Poverty," in Huber and Chalfant (eds.), *The Sociology of American Poverty*. Cambridge, MA: Schenkman Publishing Company, 1974, p. 201.
20. Edwin M. Schur, "Poverty, Violence and Crime in America," in ibid., p. 266.
21. U.S. Bureau of the Census, *Statistical Abstract of the United States: 1975* (96th edition). Washington, DC, 1975, pp. 158, 160–61.
22. Ely Chinoy, *Automobile Workers and the American Dream*. Boston: Beacon Press, 1955.
23. Charles Kadushin, "Social Class and the Experience of Ill Health," in Reinhard Bendix and Seymour Martin Lipset (eds.), *Class, Status, and Power: Social Stratification in Comparative Perspective*. New York: Free Press/Macmillan, 1966, pp. 407–11.
24. Herman E. Daly, *Essays Toward a Steady-State Economy*. Cidoc Cuaderno, No. 70, 1972.
25. Bayard Rustin, "No Growth Has To Mean Less Is Less," *The New York Times Magazine* (May 2, 1976).
26. Daniel Bell, *The Coming of Post-Industrial Society*. New York: Basic Books, 1973; and *The Cultural Contradictions of Capitalism*. New York: Basic Books, 1976.
27. Mark Kelman, "The Social Costs of Inequality," *Dissent*, Vol. XX, No. 3 (Summer 1973), pp. 291–98.
28. Gunnar Myrdal, "On the Equality Issue in World Development," *World Issues*, Vol. 1, No. 1 (Oct.-Nov. 1976), p. 5.
29. Norman Girvan, "Economic Nationalism," *Daedalus*, Vol. 104, No. 4 (Fall 1975), pp. 146–58.
30. Nathan Keyfitz, "World Resources and the World Middle Class," *Scientific American*, Vol. 235, No. 1 (July 1976), pp. 28–29.
31. William Schneider, *Food, Foreign Policy, and Raw Materials Cartels*. New York: Crane, Russak & Company, 1976.
32. Brookings Institution, *Trade in Primary Commodities: Conflict or Cooperation?* Washington, DC: 1974, p. 28.
33. Hans J. Morgenthau, "World Politics and the Politics of Oil," in Gary Eppen (ed.), *Energy: The Policy Issues*. Chicago: University of Chicago Press, 1975, p. 43.

5: Three Tactics in the Theory of Development

1. Seymour Martin Lipset, *Revolution and Counterrevolution: Change and Persistence in Social Structures.* New York: Basic Books, 1968.
2. André Gunder Frank, *Latin America: Underdevelopment or Revolution.* New York and London: Monthly Review Press, 1969.
3. Bert F. Hoselitz, *Sociological Aspects of Economic Growth.* New York: Free Press/Macmillan, 1960.
4. Daniel Lerner, *The Passing of Traditional Society.* Free Press/Macmillan, 1964.
5. Edward A. Shils, "Intellectuals in the Political Development of the New States," *World Politics,* Vol. 12, No. 3 (April 1960), pp. 329–68.
6. Samuel P. Huntington, *Political Order in Changing Societies.* New Haven: Yale University Press, 1968.
7. Celso Furtado, *Development and Underdevelopment.* Berkeley and Los Angeles: University of California Press, 1964.
8. René Dumont (with Marcel Mazoyer), *Socialisms and Development.* New York: Praeger, 1973.
9. Immanuel Wallerstein, *The Modern World System: Capitalist Agriculture and the Origins of the European World-Economy in the Sixteenth Century.* New York and London: Academic Press, 1974.
10. Paul Baran, *The Political Economy of Growth.* New York: Monthly Review Press, 1957; see also, Paul Baran and Paul M. Sweezy, *Monopoly Capital.* New York: Monthly Review Press, 1966.
11. Dale L. Johnson, James D. Cockroft, and André Gunder Frank, *Dependence and Underdevelopment.* Garden City, NY: Doubleday and Company, 1972; see also, Frederick Stirton Weaver, "Capitalist Development, Empire, and Latin American Underdevelopment," *Latin American Perspectives,* Vol. 3, No. 4 (Fall 1976), pp. 17–53.

6: From Dependency to Determinism in the Development Process

1. For an analysis of this earlier period in United States-Latin American military relations, see Irving Louis Horowitz, "The Military Élites," in Seymour Martin Lipset and Aldo Solari (eds.), *Élites in Latin America.* New York and London: Oxford University Press, 1967, pp. 146–89; and "Political Legitimacy and the Institutionalization of Crisis in Latin America," *Comparative Political Studies,* Vol. 1, No. 1 (April 1968), pp. 45–69.
2. Morris Janowitz, "Preface" to *Political Participation under Military Regimes,* Henry Brienen and David Morell (eds.), Beverly Hills and London: Sage Publications, 1976, p. 8.
3. Kalman H. Silvert, *The Conflict Society: Reaction and Revolution in Latin America* (revised edition). New York: American Universities Field Staff, 1966, p. 22.

4. José Nun, "A Latin American Phenomenon: The Middle-Class Military Coup," in *Trends in Social Science Research in Latin American Studies*. Berkeley: University of California Institute of International Studies, 1965, pp. 55–99. See also, José Nun, "The Middle-Class Military Coup Revisited," in Abraham F. Lowenthal (ed.), *Armies and Politics in Latin America*. New York and London: Holmes and Meier, 1976, pp. 49–86.
5. Fidel Castro Ruz, "Our Armed Forces Are Firmly Linked to the People, to the Revolutionary State and to Their Vanguard Party," Speech at the Conclusion of the 15th Anniversary of the Triumph of the Revolutionary Military Exercises. Havana, Cuba: Political Editions, 1974, pp. 9–12.
6. Charles C. Moskos, Jr., "Proposal for a Study on the Military Factor in Western Civilization" (mimeographed), 1976; and "The Military," *Annual Review of Sociology*, Vol. 2. Detroit: Gale Research Co., 1976, pp. 55–77.
7. R. D. McKinlay and A. S. Cohan, "Performance and Instability in Military and Nonmilitary Regime Systems," *American Political Science Review*, Vol. 70, No. 3 (September 1976), pp. 856–58.
8. James A. Barber, Jr., "The Military Services and American Society: Relationships and Attitudes," in Stephen E. Ambrose and James A. Barber, Jr. (eds.), *The Military and American Society*. New York: Free Press, 1972, p. 301.
9. John Erickson, "The Origins of the Red Army," in Richard Pipes (ed.), *Revolutionary Russia*. Cambridge, MA: Harvard University Press, 1968, pp. 286–325.
10. Celso Furtado, "The Post-1964 Brazilian 'Model' of Development," *Studies in Comparative International Development*, Vol. 8, No. 2 (Summer 1973), pp. 115–27.
11. Riordan Roett, *Brazil: Politics in a Patrimonial Society* (revised edition). New York: Praeger/Holt, Rinehart & Winston, 1978.
12. Alfred Stepan, *The State and Society: Peru in Comparative Perspective*. Princeton: Princeton University Press, 1978, pp. 301–16.
13. Manuel D. Ornellas, "El 'peruanismo' y sus lecciones," *Carta Política*, No. 34 (August 1976), p. 38.
14. Juan de Onís, "Slayings Plague Argentine Junta," *New York Times*, August 21, 1976; and "Bolivia, Five Years Under Military, Takes a Step Toward Modernity," *New York Times*, December 8, 1976.
15. *Chile Economic News* is a publication of Corporación de Fomento de la Producción (CORFO). It provides a careful monitoring of economic indicators on a monthly basis. Indeed, it offers strong evidence for the capacity of state capitalism, operating under a bureaucratic-military blanket, to move from dependency to determinism. See whole numbers 106-10 (January-May 1980) for summaries, updates, and projections on banking, investment, and commercial and industrial trends in the Chilean economy.

16. Andrés Tarnowski, "Post-Perón Argentina Is Still Chaotic," *New York Times*, July 25, 1976; and Juan de Onís, "Argentina Is Making Economic Comeback," *New York Times*, October 3, 1976.
17. Neuma Aguiar, *The Structure of Brazilian Development*. New Brunswick, NJ: Transaction Books, 1979, pp. 1–16.
18. Barry Ames, "Rhetoric and Reality in a Militarized Regime: Brazil Since 1964," in Abraham F. Lowenthal (ed.), *Armies and Politics in Latin America*, pp. 261–90.
19. Denis Gilbert, "The End of the Peruvian Revolution: A Class Analysis," *Studies in Comparative International Development*, Vol. XV, No. 1 (Spring 1980), pp. 15–38.
20. Juan de Onís, "Peruvian Giving Away from Left," *New York Times*, December 24, 1976; and Joanne Omang, "Peru Soft-Pedals Revolution," *Washington Post*, September 19, 1976.
21. Thomas E. Weil et al., "Area Handbook for Paraguay," *Foreign Area Studies of The American University*. Washington, DC: U.S. Government Printing Office, 1972; and "Dirección General de Estadística y Censos, Censo nacional de población y viviendas, República del Paraguay." Asunción, Paraguay: La Dirección, 1975.
22. Richard Arens (ed.), *Genocide in Paraguay*. Philadelphia: Temple University Press, 1977, pp. 98–101.
23. Juan J. Linz, "The Future of an Authoritarian Situation or the Institutionalization of an Authoritarian Regime: The Case of Brazil," in Alfred Stepan (ed.), *Authoritarian Brazil: Origins, Policies, and Future*. New Haven and London: Yale University Press, 1973, pp. 233–52.
24. Jonathan Kandell, "Grip of Latin Military Squeezes Leftists Out," *New York Times*, July 24, 1976; Joanne Omang, "Argentine Leader Reported Trying To Curb Death Squads," *Washington Post*, September 19, 1976; and "Government in Peru Stepping Up Its Efforts To Halt Subversion and Increase Production," *New York Times*, August 29, 1976.
25. Mort Rosenblum, "Terror in Argentina," *New York Review of Books*, Vol. 23, No. 17 (October 1976), pp. 26–27.
26. Stanley G. Payne, *Politics and the Military in Modern Spain*. Stanford, CA: Stanford University Press, 1967.
27. Ben S. Stephansky and Robert J. Alexander, "Report of Commission on Enquiry into Human Rights in Paraguay" (mimeographed). New York: International League for Human Rights, 1976, pp. 2–5.
28. Alfred Stepan, *The Military in Politics: Changing Patterns in Brazil*. Princeton: Princeton University Press, 1971, pp. 267–71. Stepan's notion that the mode of military involvement may well shift from that been the first serious expression of my thesis on the movement from dependence to determinism.
29. Cynthia Enloe, "Ethnicity and Militarization Factors Shaping the Roles

of Police in Third World Nations," *Studies in Comparative International Development*, Vol. 12, No. 3 (Autumn 1976), pp. 326–39.

30. Joanne Omang, "Where Might Makes Right," *Washington Post*, August 29, 1976.
of system-maintenance to that of system-transformation may well have
31. Robin Luckham, "The Military, Militarism and Dependence in the Third World," in C. H. Enloe and U. Semin-Panzer (eds.), *The Military, the Police, and Domestic Order: British and Third World Experiences*. London: Richardson Institute for Conflict and Peace Research, 1976, p. 196.
32. Mark Falcoff, "Our Latin American Hairshirt," *Commentary*, Vol. 62, No. 4 (October 1976), p. 61.
33. Cyril E. Black, "Military Leadership and National Development," in Davic Mac Isaac (ed.), *Proceedings of the Fifth Military History Symposium*. Boulder, CO: United States Air Force Academy, 1975, pp. 16–35.
34. Fernando Henrique Cardoso, "Associated Dependent Development: Theoretical and Practical Implications," in Alfred Stepan (ed.), *Authoritarian Brazil*. New Haven: Yale University Press, 1973, pp. 142–78.
35. Edy Kaufman, *Uruguay in Transition: From Civilian to Military Rule*. New Brunswick, NJ: Transaction Books, 1979.
36. Robert C. Sellers, *Armed Forces of the World: A Reference Handbook*, 4th ed. New York: Praeger, 1977, pp. 169–70.

7: Globalizing War and Integrating World Systems

1. The literature on nuclear conflict is exhaustive and extensive. A sense of differences in interpretation is suggested by the following classic texts: Edward Teller, *The Legacy of Hiroshima*. London: Macmillan, 1962; C. Wright Mills, *The Causes of World War Three*. New York: Simon and Schuster, 1958; and Herman Kahn, *On Thermonuclear War*. Princeton: Princeton University Press, 1960.
2. W. Scott Thompson, *Power Projection: A Net Assessment of United States and Soviet Capabilities*. (Agenda Paper for National Strategy Information Center.) New Brunswick, NJ: Transaction Books, 1980.
3. Irving Louis Horowitz, *The War Game: Studies of the New Civilian Militarists*. New York: Ballantine Books/Random House, 1963.
4. Jerry F. Hough, "Why the Russians Invaded," *The Nation*, Vol. 230, No. 8 (March 1, 1980).
5. Richard Pipes, "Soviet Global Strategy," *Commentary*, Vol. 69, No. 4 (April 1980), pp. 34–35.
6. Edward N. Luttwak, "After Afghanistan, What?" *Commentary*, Vol. 69, No. 4 (April 1980), p. 45.
7. Samual Pisar, *Coexistence and Commerce: Guidelines for Transactions*

Between East and West. New York: McGraw Hill Book Company, 1970, pp. 94–114.

8. Edward N. Luttwak, *Strategy and Politics: Collected Essays*. New Brunswick, NJ: Transaction Books, 1980.
9. The information of current deaths through violent conflicts was issued by the Center for Defense Information, Washington, DC, in 1979. For a moving essay on the subject, see Lewis H. Lapham, "Voices Prophesying War," *Harper's*, Vol. 260, Whole No. 1559 (April 1980), pp. 12–16.

8: State Power and Military Nationalism

1. André Gunder Frank, *Capitalism and Underdevelopment in Latin America*. New York: Monthly Review Press, 1969; and André Gunder Frank, *Latin America: Underdevelopment or Revolution?* New York and London: Monthly Review Press, 1970.
2. Fernando Henrique Cardoso and Enzo Faletto, *Dependencia y desarrollo en America Latina*. Mexico, D.F.: Siglo Veintiuno, 1970.
3. Philippe C. Schmitter, "The 'Portugalization' of Brazil?" in Alfred Stepan (ed.), *Authoritarian Brazil: Origins, Policies, and Future*. New Haven and London: Yale University Press, 1973, pp. 179–232.
4. Nicos Poulantzas, *Political Power and Social Classes*, translated by Timothy O'Hagan. London and New York: New Left Books and Humanities Publishers, 1973, p. 247.
5. Ibid., p. 262.
6. Ibid., p. 115; see also by the same author, "The Problem of the Capitalist State," *New Left Review*, Whole No. 58 (Nov./Dec. 1970), pp. 67–78.
7. Ralph Miliband, "Poulantzas and the Capitalist State," *New Left Review*, Whole No. 82 (Nov./Dec. 1973), pp. 87–88.
8. Samuel P. Huntington, *Political Order in Changing Societies*. New Haven and London: Yale University Press, 1968, p. 20.
9. Ibid., p. 91.
10. Ralph Miliband, *The State in Capitalist Society*. London: George Allen & Unwin, 1969, p. 187.
11. Reinhard Bendix, "Social Stratification and the Political Community," in Reinhard Bendix and Seymour Martin Lipset (eds.), *Class, Status and Power: Social Stratification in Comparative Perspective*. New York: Free Press/Macmillan, 1966, pp. 73–86.
12. Edward A. Shils, "The Military in the Political Development of the New States," in John J. Johnson (ed.), *The Role of the Military in Underdeveloped Countries*. Princeton: Princeton University Press, 1962, Chapter 1.
13. Celso Furtado, "The Post-1964 Brazilian Model of Development," *Studies in Comparative International Development*, Vol. 8, No. 2 (Summer 1973), pp. 117–18; and Irving Louis Horowitz, *Three*

Worlds of Development: The Theory and Practice of International Stratification (second edition). New York and London: Oxford University Press, 1972, pp. 284–314.

14. Edward Feit, "Pen, Sword, and People: Military Regimes in the Formation of Political Institutions," *World Politics*, Vol. 25 (January 1973), pp. 251–73.
15. Aleksandr I. Solzhenitsyn, *The Gulag Archipelago: 1918–1956*, translated by Thomas P. Whitney. New York: Harper & Row, 1974, pp. 24–92.
16. Schmitter, p. 179.
17. Hélio Jaguaribe, *Economic and Political Development: A Theoretical Approach and a Brazilian Case Study*. Cambridge, MA: Harvard University Press, 1968, pp. 191–92.
18. Werner Baer, Isaac Kerstenetzky, and Aníbal Villela, "The Changing Role of the State in the Brazilian Economy," *World Development*, Vol. 1, No. 11 (November 1973), pp. 23–24; see also Werner Baer, "The Brazilian Boom of 1968–72: An Explanation and Interpretation," *World Development*, Vol. 1, No. 8 (August 1973), pp. 1–15.
19. Peter B. Evans, "The Military, the Mutinationals and the 'Miracle': The Political Economy of the 'Brazilian Model' of Development," *Studies in Comparative International Development*, Vol. 9, No. 3 (Fall 1974), pp. 26–45.
20. Juan J. Linz, "The Future of an Authoritarian Situation or the Institutionalization of an Authoritarian Regime: The Case of Brazil," in Alfred Stepan (ed.), *Authoritarian Brazil: Origins, Policies, and Future*, pp. 179–232.
21. Carlos Estevam Martins, "Brazil and the United States from the 1960s to the 1970s," in Julio Cotler and Richard Fagen (eds.), *Latin America and The United States: The Changing Political Realities*. Stanford: Stanford University Press, 1974, pp. 269–301.
22. Luigi R. Einaudi, "Revolution from Within? Military Rule in Peru Since 1968," *Studies in Comparative International Development*, Vol. 8, No. 1 (Spring 1973), pp. 71–87.
23. See Juan de Onís, "Peru's New Chief Won by Shrewd Use of Military Issue," *New York Times*. May 21, 1980.
24. Pablo González Casanova, *Democracy in Mexico*, translated by Danielle Salti. New York and London: Oxford University Press, 1970, pp. 11–30; see also Nora Hamilton, "Dependent Capitalism and the State: The Case of Mexico." Abstract and summary of a dissertation in the Department of Sociology, The University of Wisconsin (Madison, 1972).
25. Bo Anderson and James D. Cockroft, "Control and Cooptation in Mexican Politics," in Irving Louis Horowitz, Josué de Castro, John Gerassi (eds.), *Latin American Radicalism: A Documentary Report on Left and Nationalist Movements*. New York and London: Random House, 1969, pp. 366–89.

26. Shils, "The Military in the Political Development of the New States," pp. 7–67.
27. Alfred Stepan, *The Military in Politics: Changing Patterns in Brazil*, pp. 267–71.
28. Luis Echeverría, Speech of February 18, 1974; reported in the *Los Angeles Times* (February 19, 1974).
29. Henry Kissinger, "Good Partner Policy for the Americas," *Society*, Vol. 11, No. 6 (Sept./Oct. 1974), pp. 16–22; see also Richard R. Fagen, "The 'New Dialogue' on Latin America," ibid., pp. 17–30.
30. This chapter incorporates material from "State Power and Military Nationalism in Latin America" [with E. K. Trimberger], *Comparative Politics*, Vol. 8, No. 2 (January 1976), pp. 223–44.

9: Legitimacy and Illegitimacy in Third World Regimes

1. Irving Louis Horowitz, "Party Charisma: Practices and Principles," "Ideological and Institutional Bases of Party Charisma," and "Intra-Country Variations in Party Charisma," *Three Worlds of Development: The Theory and Practice of International Stratification* (first edition). New York and London: Oxford University Press, 1966, pp. 225–53.
2. Irving Louis Horowitz, "Political Legitimacy and the Institutionalization of Crisis," *Comparative Political Studies*, Vol. 1, No. 1 (1968), pp. 45–69; and "The Norm of Illegitimacy: The Political Sociology of Latin America," in I. L. Horowitz, J. de Castro, and J. Gerassi (eds.), *Latin American Radicalism: A Documentary Report on Left and Nationalist Movements*. New York: Random House; London: Jonathan Cape, 1969, pp. 3–29.
3. Alan Wolfe, *The Limits of Legitimacy: Political Contradictions of Contemporary Capitalism*. New York: Free Press/Macmillan, 1977.
4. Julius Jacobson, *Soviet Communism and the Soviet Vision*. New Brunswick, NJ: Transaction Books/E. P. Dutton, 1972.
5. David C. Schwartz, Joel Aberbach, Ada W. Finifter, et al., "Political Alienation in America," A special supplement to *Transaction/SOCIETY*, Vol. 13, No. 5 (July/August 1976), pp. 18–53.
6. Bogdan Denis Denitch, *The Legitimation of a Revolution: The Yugoslav Case*. New Haven and London: Yale University Press, 1976, pp. 41–48.
7. Leslie F. S. Upton, *The United Empire Loyalists: Men and Myths*. Toronto: Copp Clark Publishing Company, 1967, p. 7.
8. David Lane, *Politics and Society in the USSR*. New York: Random House, 1971, pp. 59–61.
9. Clifford Geertz (ed.), *Old Societies and New States: The Quest for Modernity in Asia and Africa*. New York: Free Press/Macmillan, 1963, pp. 105–57.
10. Georges Balandier, *Political Anthropology*. New York: Pantheon Books (A Division of Random House), 1970.

11. Frederick N. Nunn, *The Military and Chilean History: Essays on Civil/Military Relations, 1810–1973*. Albuquerque: University of New Mexico Press, 1976.
12. Edy Kaufman, *Uruguay in Transition: From Civil to Military Rule*. New Brunswick, NJ: Transaction Books, 1978.
13. Abraham F. Lowenthal (ed.), *Armies and Politics in Latin America*. New York and London: Holmes and Meier Publishers, 1976.
14. Alfred Stepan, *The Military in Politics: Changing Patterns in Brazil*.
15. Martin C. Needler, "The Closeness of Elections in Latin America," *Latin American Research Review*, Vol. XII, No. 1, pp. 115–21.

10: Military Origins of Third World Dictatorship and Democracy

1. Walter Laqueur, "Fascism: The Second Coming," *Commentary*, Vol. 61, No. 2 (February 1976), pp. 61–62. See also *The Political Psychology of Appeasement*. New Brunswick, NJ: Transaction Books, 1980, pp. 189–210.
2. Noam Chomsky, *The Washington Connection and Third World Fascism: The Political Economy of Human Rights*. Boston: South End Press, 1979.
3. C. B. Macpherson, *The Real World of Democracy*. London: Oxford University Press, 1966, pp. 23–34.
4. W. W. Rostow and Richard W. Hatch, *An American Policy in Asia*. Cambridge, MA: MIT Press, and New York: Wiley, 1955; see also *The Diffusion of Power: An Essay in Recent History*. New York: Macmillan, 1972.
5. Seymour Martin Lipset, *Revolution and Counterrevolution: Change and Persistence in Social Structures*. New York: Basic Books, 1968.
6. Daniel Lerner, *The Passing of Traditional Society: Modernizing the Middle East*. Glencoe, IL: The Free Press, 1958.
7. Wilber L. Schramm, *Communication in Modern Society*. Urbana: University of Illinois Press, 1948; and *Mass Media and National Development: The Role of Information in the Developing Countries*. Stanford: Stanford University Press, 1964.
8. Marta Panaia and Ricardo Lesser, "Las estrategias militares frente al proceso de industrializacion (1943–1947), *Estudios Sobre los orígenes del peronismo* /2. Buenos Aires: Siglo XXI Argentina Editores, 1973. pp. 83–164.
9. Jean Herskovits, "Democracy in Nigeria," *Foreign Affairs*, Vol. 58, No. 2 (Winter 1979/80), pp. 314–35.
10. Kemal Karpat, *Turkey's Politics: The Transition to a Multi-Party System*. Princeton: Princeton University Press, 1959; and Feroz Ahmad, *The Young Turks*. London: Oxford University Press, 1969.
11. Gamal Abdel Nasser, *Egypt's Liberation: Philosophy of the Revolution*. Washington, DC: Public Affairs Press, 1955, pp. 24–40; see also

Nissim Rejwan, *Nasserist Ideology: Its Exponents and Critics*. New York: John Wiley & Sons, and New Brunswick, NJ: Transaction Books, 1974.

12. Moshe Lissak, *Military Roles in Modernization*. Beverly Hills and London: Sage Publications, 1976, pp. 247–48.
13. Eliezar Be'eri, *Army Officers in Arab Politics and Society*. New York: Praeger/Holt, Rinehart & Winston, 1970, pp. 351–59.
14. See C. Wright Mills, *Power, Politics and People*, edited by Irving Louis Horowitz. New York: Oxford University Press, 1963, esp. pp. 23–38; 110–39; 305–23.
15. Otto Nathan, *The Nazi Economic System: Germany's Mobilization for War*. Durham, NC: Duke University Press, 1944, pp. 365–77.
16. John Kenneth Galbraith, *The New Industrial State*. Boston: Houghton Mifflin, 1967, pp. 304–23.
17. Juan J. Linz, "The Future of an Authoritarian Situation or the Institutionalization of an Authoritarian Regime: The Case of Brazil," in Alfred Stepan (ed.), *Authoritarian Brazil*, p. 251.

11: Capitalism, Communism, and Multinationalism

1. Charles Kindleberger, *American Business Abroad*. New Haven: Yale University Press, 1969; and *The International Corporation*. Cambridge, MA: MIT Press, 1970.
2. Michael Tanzer, *The Political Economy of International Oil and the Underdeveloped Countries*. Boston: Beacon Press, 1969.
3. Dennis M. Ray, "Corporations and American Foreign Relations," *Annals of the American Academy of Political and Social Science*, No. 403 (September 1972), pp. 80–92.
4. Thomas Perry Thornton (ed.), *The Third World in Soviet Perspective*. Princeton: Princeton University Press, 1964.
5. Samuel Pisar, *Coexistence and Commerce: Guidelines for Transactions Between East and West*. New York: McGraw-Hill, 1979, pp. 352–53.
6. Raymond Vernon, *Sovereignty at Bay: The Multinational Spread of U.S. Enterprises*. New York: Basic Books, 1971, p. 259.
7. Harlan H. Vinnedge, "Another Rum Deal with Russia." *The Nation*, Vol. 215, No. 18 (December 4, 1972), pp. 558–59.
8. Lester R. Brown, *World without Borders*. New York: Random House, 1972, pp. 228–29.
9. Jean Boddewyn and Étienne F. Cracco, "The Political Game in World Business," *Harvard Business Review* (Jan./Feb. 1972); and David H. Blake (ed.), "The Multinational Corporation," *Annals of the American Academy of Political and Social Science* (September 1972), pp. 1–247.
10. Gus Tyler, "Multinational Corporations vs. Nations," *Current*, No. 143 (September 1972), pp. 54–62.
11. John Gennard, "British Trade Union Response to the Multinational Corporation," *Looking Ahead*, Vol. 20, No. 2 (March 1972), pp. 1–8.

12. Charles Levinson, *Capital, Inflation and the Multinationals.* London: George Allen & Unwin, 1971, pp: 209–10.
13. Zbigniew Brzezinski and Samuel P. Huntington, *Political Power: USA/USSR.* New York: Viking Press, 1964, pp. 429–31.
14. Edward Weisband and Thomas M. Franck, *Word Politics: Verbal Strategy among the Superpowers.* New York and London: Oxford University Press, 1971.
15. David Burtis, Farid Lavipour, Steven Ricciardi, and Karl P. Sauvant, *Multinational Corporation-Nation State Interaction.* Philadelphia: Foreign Policy Research Institute, 1971.
16. Louis Turner, *Invisible Empires.* New York: Harcourt Brace Jovanovich, 1971, pp. 185–87.
17. George F. Kennan, "Interview with George F. Kennan," *Foreign Policy,* No. 7 (Summer 1972), pp. 5–21.

12: Human Rights and the Developmental Process

1. Richard McKeon and Stein Rokkan, *Democracy in a World of Tensions: A Symposium Prepared by UNESCO.* Chicago: University of Chicago Press, 1951.
2. See Sidney Hook, "Absolutism and Human Rights," in Sidney Morgenbesser, Patric Suppes, and Morton White (eds.), *Philosophy, Science and Method: Essays in Honor of Ernest Nagel.* New York: St. Martin's Press, 1969, pp. 382–99.
3. Robert Nisbet, *Twilight of Authority.* New York: Oxford University Press, 1975, pp. 160–62.
4. August 11, 1978.
5. The most impressive application of social indicators to issues of human rights on a world scale is contained in *Yearbook on Human Rights for 1973–74.* New York and Paris: United Nations, 1977, pp. 269–77.

13: Bureaucracy, Administration, and State Power

1. The two most prescient sources on defining the nature of post-industrialism are Daniel Bell, *The Coming of Post-Industrial Society: A Venture in Social Forecasting.* New York: Basic Books, 1973; and *The Cultural Contradiction of Capitalism.* New York: Basic Books, 1976. For an earlier evaluation of mine on Bell's latter work, see Irving Louis Horowitz, "A Funeral Pyre for America," *Worldview,* Vol. 19, No. 11 (November 1976).
2. Harlan Cleveland, "The American Public Executive: New Functions, New Style, New Purpose," in James C. Charlesworth (ed.), *Theory and Practice of Public Administration: Scope, Objectives, and Methods.* Philadelphia: The American Academy of Political and Social Science, 1968, pp. 168–78.
3. H. W. Singer, "Multinational Corporations and Technology Transfer,"

The Strategy of International Development: Essays in the Economics of Backwardness. White Plains, NY: International Arts and Science Press, 1975, pp. 208–33.

4. B. Bruce-Briggs, "Enumerating the New Class," *The New Class?* New Brunswick, NJ: Transaction Books, 1979, pp. 217–25.
5. Max Weber, "Bureaucracy," in Hans Gerth and C. Wright Mills (eds.), *From Max Weber: Essays in Sociology.* New York: Oxford University Press, 1946, pp. 196–244.
6. See James G. March and Herbert Simon, "The Theory of Organizational Equilibrium," *Organizations.* New York: John Wiley, 1958, pp. 84–108; and also Dorwin Cartwright, "Influence, Lêadership, Control," in James G. March (ed.), *Handbook of Organizations.* Chicago: Rand McNally, 1964, pp. 1–47.
7. Irving Louis Horowitz, "On the Expansion of New Theories and the Withering Away of Old Classes," *Transaction/SOCIETY*, Vol. 16, No. 2 (Jan./Feb. 1979), pp. 55–62.
8. Irving Louis Horowitz, "Methods and Strategies in Evaluating Equity Research," *Social Indicators Research*, Vol. 16, No. 1 (Jan. 1979), pp. 1–22.
9. Irving Louis Horowitz, "Social Welfare, State Power, and the Limit of Equity," in Joseph Grunfeld (ed.), *Growth in a Finite World.* Philadelphia: The Franklin Institute Press, 1979, pp. 21–35.
10. Charles E. Lindblom, "The Science of Muddling Through," *Public Administration Review*, Vol. 19 (1959), pp. 79–88, and "Policy Analysis," *American Economic Review*, Vol. 48 (1958), pp. 298–99. See also his more recent work, *Politics and Markets: The World's Political-Economic Systems.* New York: Basic Books, 1977, esp. pp. 119–43; and Bill G. Schumacher, *Computer Dynamics in Public Administration.* New York: Spartan Books, 1967, pp. 163–71.
11. Dwight Waldo, "Scope of the Theory of Public Administration," *Theory and Practice of Public Administration.* Philadelphia: American Academy of Political and Social Science, 1968, p. 23.
12. Samuel B. Bacharach, "What's Public Administration? An Examination of Basic Textbooks," *Administrative Science Quarterly*, Vol. 21, No. 2 (June 1976), pp. 346–51.
13. Michel Crozier, *The Bureaucratic Phenomenon.* London: Tavistock, 1964; and more recently, *The Stalled Society.* New York: Viking Publishers, 1973.
14. Nicos Poulantzas, *State, Power, Socialism.* London: New Left Books, 1978; and his earlier, albeit less decisive, enunciation of the same theme, *Political Power and Social Classes.* London: New Left Books, 1973.
15. Ernest Mandel, *Late Capitalism.* London: New Left Books, 1975, pp. 405–7.
16. Poulantzas, *State, Power, Socialism.* Pp. 127–39.
17. Michel Crozier, *La Société bloquée.* Paris: Éditions du Seuil, 1970, p. 20; see the discussion of this in Michael Rose, *Servants of Post-*

Industrial Power?, Sociologie du Travail in Modern France. White Plains, NY: M. E. Sharpe, Inc., pp. 113–27.

18. Poulantzas, *State, Power, Socialism*. Pp. 251–65.
19. Jean-Jacques Servan-Schreiber, *The Radical Alternative*. New York: W. W. Norton, Inc., 1971, pp. 59–61.
20. Aaron Wildavsky, *Budgeting: A Comparative Theory of Budgetary Processes*. Boston: Little Brown & Company, 1975, pp. 155–57.
21. Joseph LaPalombara, *Bureaucracy and Political Development*. Princeton: Princeton University Press, 1963, pp. 48–55.
22. Gordon Tullock, *The Politics of Bureaucracy*. Washington, DC: Public Affairs Press, 1965, p. 181.
23. William Howard Gammon and Lowell H. Hattery, "Managing the Impact of Computers on the Federal Government," *The Bureaucrat*, Vol. 7, No. 2 (Summer 1978), pp. 18–19.
24. Ron Johnson and Philip Gummett, *Directing Technology: Policies for Promotion and Control*. New York: St. Martin's Press, 1979, pp. 13–14.
25. Alan Peacock, "Public Expenditure Growth in Post-Industrial Society," in Bo Gustafsson (ed.), *Post-Industrial Society*. New York: St. Martin's Press, 1979, pp. 91–95.
26. Manfred Stanley, *The Technological Conscience: Survival and Dignity in an Age of Expertise*. New York: Free Press/Macmillan, 1978, pp. 251–53.
27. David Apter, *Choice and the Politics of Allocation*. New Haven and London: Yale University Press, 1979, pp. 128–54; and Edward R. Tufte, *Political Control of the Economy*. Princeton: Princeton University Press, 1978, pp. 110–45.
28. For strongly contrasting definitions of democracy see Dorothy Pickles, *Democracy*. New York: Basic Books, 1970, pp. 9–28, 169–82; and C. B. Macpherson, *Democratic Theory*. London: Oxford University Press, 1973, esp. pp. 3–23, 29–76.

14: Social Welfare, State Power, and Limits to Equity

1. Richard I. Curtin, "Perceptions of Distributional Equity: Their Economic Bases and Consequences." Ph.D. dissertation, University of Michigan, 1976.
2. Samuel P. Huntington, "The Democratic Distemper," *The Public Interest*. Whole No. 41 (Fall 1975), pp. 17–18; see also Michel Crozier, Samuel P. Huntington, and Joji Watanuki, *The Crisis of Democracy*. New York: New York University Press, 1975.
3. Edward F. Renshaw, *The End of Progress: Adjusting to a No-Growth Economy*. North Scituate, MA: Duxbury Press, 1976, pp. 7–10.
4. E. J. Mishan, *Economics for Social Decisions: Elements of Cost-Benefit Analysis*. New York: Praeger, 1972; also *Cost-Benefit Analysis* (second edition). New York: Praeger, 1976, pp. 416–49.
5. Herman Kahn and Anthony J. Wiener, *Year Two Thousand*. New

York: Macmillan, 1967; and also Herman Kahn et al., *The Next Two Hundred Years*. Caldwell, NJ: Morrow, 1976.

6. John W. Bennett, "Anticipation, Adaptation, and the Concept of Culture in Anthropology," *Science*, Vol. 192, No. 4242 (May 28, 1976), pp. 847–53.
7. F. Thomas Juster, "The Recovery Gathers Momentum," *Economic Outlook. USA*, Vol. 3, No. 2 (Spring 1976), pp. 23–25.
8. Paul Neurath, "Zwischen Pesimismus und Optimismus," in Wolf Fruhauf (ed.), *Wissenschaft und Weltbild (Festschrift fur Hertha Firnberg)*. Vienna: Europaverlag, 289–312.
9. André van Dam, "A Simpler Life for the Advanced Countries," *Progress International*, Vol. 2 (November 1975), pp. 10–12.
10. Eileen Shanahan, "Income Distribution Found Little Changed Since War," *New York Times* (February 2, 1974), p. 10.
11. David Donnison, "Equality," *New Society*, Vol. 34, Whole No. 685 (November 20, 1975), p. 423.
12. Karl W. Deutsch, "On Inequality and Limited Growth: Some World Political Effects," *International Studies Quarterly*, Vol. 19, No. 4 (December 1975), pp. 381–89.
13. Jay W. Forrester, *World Dynamics*. Cambridge, MA: Wright-Allen Publishers, 1971; and "Limits to Growth Revisited," *Journal of the Franklin Institute*, Vol. 300, No. 2 (August 1975), pp. 107–11.
14. Samuel P. Huntington, "The Democratic Distemper," *The Public Interest*, Whole No. 41 (Fall 1975), p. 37.
15. Edgar K. Browning, "How Much More Equality Can We Afford?" *The Public Interest*, Whole No. 43 (Spring 1976), pp. 90–110.

15: Democracy and Development: Policy Perspectives

1. See Ernest W. Lefever (ed.), *Will Capitalism Survive?* Washington, DC: Ethics and Public Policy Center of Georgetown University, 1979, esp. pp. 3–14, 54–57.
2. Ali Maruzi, "The New Interdependence," in Guy F. Erb and Valeriana Kallab (eds.), *Beyond Dependency: The Developing World Speaks Out*. New York: Praeger, 1975, pp. 38–54.
3. Immanuel Wallerstein, *The Modern World-System: Capitalist Agriculture and the Origins of the European World-Economy in the Sixteenth Century*. New York and London: Academic Press, 1974, pp. 350–57.
4. Guillermo O'Donnell, "Tensions in the Bureaucratic-Authoritarian State and the Question of Democracy," in David Collier (ed.), *The New Authoritarianism in Latin America*. Princeton: Princeton University Press, 1979, esp. pp. 314–17.
5. Carl Stone, "Decolonization and the Caribbean State System," paper delivered at the Conference on Decolonization, Development, and Democracy in the Caribbean. Center for Inter-American Relations, May 5–6, 1978, esp. pp. 8–13.

6. Robert A. Dahl and Edward R. Tufte, *Size and Democracy*. Stanford: Stanford University Press, 1973, esp. pp. 110–17.
7. Edwin Jones, "Bureaucracy as a Problem-Solving Mechanism in Small States," in Vaughan A. Lewis (ed.), *Size, Self-Determination and International Relations: The Caribbean*. Jamaica: Institute of Social and Economic Research, University of the West Indies, 1976, pp. 75–76.
8. Gino Germani, *Marginality*. New Brunswick, NJ, and London: Transaction Books, 1980, pp. 42–43.
9. Samuel P. Huntington, "The Democratic Distemper," *The Public Interest*, pp. 9–38.
10. Irving Louis Horowitz, "Social Planning and Social Science: Historical Continuities and Comparative Discontinuities," in Robert W. Burchell and George Sternlieb (eds.), *Planning Theory in the 1980s: A Search for Future Directions*. New Brunswick, NJ: Center for Urban Policy Research, 1978, pp. 41–68.

16: The Limits of Modernity

1. In perhaps the most forceful and persuasive defense of Khomeini's regime, Eric Rouleau notes that "rare were those who suggested that modernity is not necessarily synonymous with progress or well-being, or that the concepts of the economic development current in the West —where quick material gain is often the only valid criterion—does not necessarily correspond to the true needs and interests of developing nations. Rarer still were those who pointed out the pitfalls of labelling an entire people fanatics simply because they were virtually unanimous in expressing their will." Far from being critical of this attack on modernism, Rouleau sees it as part of a long line extending from Savonarola, Calvin, and Cromwell. See Eric Rouleau, "Khomeini's Iran," *Foreign Affairs*, Vol. 59, No. 1 (Fall 1980), pp. 1–20.
2. In summarizing the results of a recent conference on conditions for change in the climate of opinion, Markovits and Deutsch pointed out that "very rapid change may favor extremes in human behavior, both in fear and in daring." They go on to note that decisions reached under certain conditions force us to reveal our values. This might be viewed as an essential component to the extreme swing between over-complexity and over-simplicity. See Andrei S. Markovits and Karl W. Deutsch, *Fear of Science—Trust in Science*. Cambridge, MA: Oelgeschlager, Gunn & Hain, 1980, pp. 230–34.
3. A much more sanguine view of this revolt against modernization has been registered by Seymour Martin Lipset and Earl Raab in "The Election and the Evangelicals," *Commentary*, Vol. 71, No. 3 (March 1981), pp. 25–31. They see higher education as well as urbanization limiting such an impact. They view the power of the evangelical movement as reflective of a mood and a sentiment favoring a fight

against inflation and a refurbishing of industry. But the issue seems less one of backlash moralistic politics than a basic reconsideration by large numbers of people through the advanced industrial belt of the meaning and purpose of modernization. And it seems dangerously mechanistic to view urbanization and education as safeguards against such a traditionalist ideology, when it is precisely in quite ultra-modern centers such as Dallas and Los Angeles that one finds the "new evangelicals" strongest—and often centered in seats of higher learning at that.

4. It is interesting to note how, until most recently, the best thinking on this subject assumed that on the poliical level, modernizaion in the Third World meant a transition from traditional authoritarian systems to more effective tyrannies. "It has now become plain that this transition at the functional level, from authoritarianism to totalitarianism, has also signified a profound shift in ideology from modernization to anti-modernization, or traditionalism." See Walter Laqueur, "Fascism: The Second Coming," in *The Political Psychology of Appeasement.* New Brunswick, NJ: Transaction Books, 1980, pp. 208–10.
5. On the heterogeneity of American investment abroad, see C. Fred Bergsten, Thomas Horst, and Theodore H. Moran, *American Multinationals and American Interests*. Washington, DC: The Brookings Institution, 1978, pp. 71–98.

Name Index

Subject Index